Wireless Networks

Series editor

Xuemin Sherman Shen
University of Waterloo, Waterloo, Ontario, Canada

More information about this series at http://www.springer.com/series/14180

Xiang Cheng • Rongqing Zhang
Liuqing Yang

5G-Enabled Vehicular Communications and Networking

Springer

Xiang Cheng
The State Key Laboratory of Advanced Optical Communication Systems and Networks, School of Electronics Engineering and Computing Science
Peking University
Beijing, Beijing, China

Rongqing Zhang
Department of Electrical and Computer Engineering
Colorado State University
Fort Collins
CO, USA

Liuqing Yang
Department of Electrical and Computer Engineering
Colorado State University
Fort Collins
CO, USA

ISSN 2366-1186 ISSN 2366-1445 (electronic)
Wireless Networks
ISBN 978-3-030-02175-7 ISBN 978-3-030-02176-4 (eBook)
https://doi.org/10.1007/978-3-030-02176-4

Library of Congress Control Number: 2018959756

This Springer imprint is published by the registered company Springer Nature Switzerland AG
The registered company address is: Gewerbestrasse 11, 6330 Cham, Switzerland

Preface

With vehicles growing increasingly intelligent and connected to form Internet of intelligent vehicles, and with wireless technology facing increasing demands of supporting high-mobility data traffic, these two century-old technologies will be drastically shaped by and fused into each other. In fact, even in the past century, the evolution paths of vehicles and wireless are not without crossings. For example, one key player in the wireless industry Motorola is actually named after its first internationally sold product, the car radio, whose name paired up "motorcar" with "-ola" for the meaning of "sound in motion." This early date crossing of vehicles and wireless has brought success for both industries, though radio is a rather passive and one-directional form of wireless communications. At that stage, wireless technology essentially adds fancy features to vehicles. These features are certainly nice to have, but one can certainly live without. Recently, endeavors of various companies including Google, Tesla Motors, and Baidu, together with major demonstrations in Europe, North America, Japan, and China, have brought self-driving intelligent vehicles closer to reality than ever. To date, most research and development efforts have been focused on the environment sensing and self-driving capabilities of a single vehicle. However, after several deadly accidents involving vehicles in self-driving mode, it has been realized that vehicle safety and reliability could only be ensured and enhanced via vehicular communications and networking. Additionally, with vehicles becoming more intelligent, the human drivers will be released from intense driving, turning vehicles into offices, meeting rooms, and entertainment centers. Each vehicle will then become much more data hungry. In the meantime, electric vehicles are promptly gaining popularity due to both fossil fuel scarcity and environmental concerns, calling for extensive and timely communications in support of the critical energy management.

At present, we have way passed the age of using wireless just as nice supplement for vehicles but are still under way toward the era of intimate communications and vehicle integration, where their core functionalities are fused and jointly designed and optimized. We term the current stage as vehicle-communication combination, in which wireless communications are tailored for the vehicular environments, such as the unique mobility-induced propagation channel, and the transportation-specific topology. Compared to the traditional approach of simply taking a generic wireless system and deploying in a vehicle, the present approach of judiciously adapting communications to vehicular environments is a huge improvement, but is not yet optimally suited for core vehicular functions.

Hence, from both the wireless-vehicle combination and integration perspectives, this monograph provides a comprehensive picture of 5G-enabled vehicular communications and networks (5G-VCN). Specifically, in Chap. 1, the development of intelligent vehicles (IVs) is first introduced, and the key features as well as the challenges for 5G-VCN are discussed. In Chap. 2, the vehicular channel characteristics and modeling are discussed and a new generic geometry-based stochastic modeling is proposed, which provides solid fundamental supports for the designs and improvements of 5G-VCN technologies. Then, from the wireless-vehicle combination perspective, in Chap. 3, we introduce and discuss advanced PHY techniques suitable for 5G-VCN in detail, including ICI cancellation, index modulated (IM-)OFDM, differential spatial modulation (DSM), energy harvesting (EH)-based vehicular communications, and some next-leap directions. Following the wireless-vehicle combination perspective, in Chap. 4, we further elaborate on effective MAC designs in both distributed and centralized manners targeting at 5G-VCN, including distributed congestion control, centralized resource sharing and scheduling, and centralized data dissemination. Finally, in Chap. 5, from the wireless-vehicle integration perspective, we explore and discuss some important 5G-VCN-based applications, including electric vehicles, distributed data storage, and physical layer security. In addition, autonomous driving is also discussed as the next leap for 5G-VCN-based IV application.

The primary audience of this book includes professionals and researchers relevant to the fields of vehicular communications, intelligent transportation systems (ITS), and Internet of vehicles (IoV). This monograph also provides the state of the art on vehicular communications and networking for people outside the mentioned fields who aspire to explore new interdisciplinary directions and research ideas with these fields.

We would like to thank Ms. Yiran Li, Dr. Miaowen Wen, Dr. Xia Shen, Dr. Dexin Wang, Dr. Luoyang Fang, Mr. Meng Zhang, Dr. Yuke Li, and Mr. Binbin Hu, for their inspiring collaborative contributions and discussions on the research topics presented in this monograph. Finally, we would like to thank the continued support from the National Natural Science Foundation of China under Grants 61622101 and 61571020, the Major Project from Beijing Municipal Science and Technology Commission under Grant Z181100003218007, and the National Science and Technology Major Project under Grant 2018ZX03001031.

Beijing, China Xiang Cheng

Contents

Chapter 1
Introduction to 5G-Enabled VCN

1.1 The Era of Intelligent Vehicles

Since Karl Benz invented motor cars more than 130 years ago, automobiles have undergone probably the most significant leap from their ancestors – the intelligent vehicles (IVs), particularly those with self-driving capability, are receiving unprecedented attention. A crucial driving force in this development is the wireless communication technology, which has matured since Marconi's first demonstration 120 years ago [1].

The history of intelligent vehicles has developed over the past two decades [2]. Although the first ideas were born in the 1960s, the level of maturity of the technology at that time did not allow pursuit of the original goal of implementing fully autonomous all-terrain all-weather vehicles. The first documented prototypes of automated vehicles were fielded by a few groups in the military arena in the mid 1980s. The initial stimulus that triggered these innovative ideas was provided by the military sector, which was eager to provide complete automation to its fleet of ground vehicles. It was not before the 1980s that this interest was transferred to the civil sector when governments worldwide launched the first projects, which supported a large number of researchers in these topics. The interest of the automotive industry in developing real products was triggered after feasibility studies were successfully completed and the first prototypes were demonstrated. Testing of autonomous vehicles on roads in a real environment was one of the most important milestones in the history of IVs. In 1995, the Carnegie Mellon NAVLAB group ran their "No Hands Across America" experiment [3]. They demonstrated automated steering, based solely on computer vision, over 98% of the time on a 2800 mile trip across the United States. Later in 1995 the Bundeswehr Universit at Munich (UBM), Germany fielded a vehicle that was demonstrated with a 1758 km trip from Munich to Copenhagen in Denmark and back. The vehicle was able to drive autonomously for 95% of the trip.

X. Cheng et al., *5G-Enabled Vehicular Communications and Networking*, Wireless Networks, https://doi.org/10.1007/978-3-030-02176-4_1

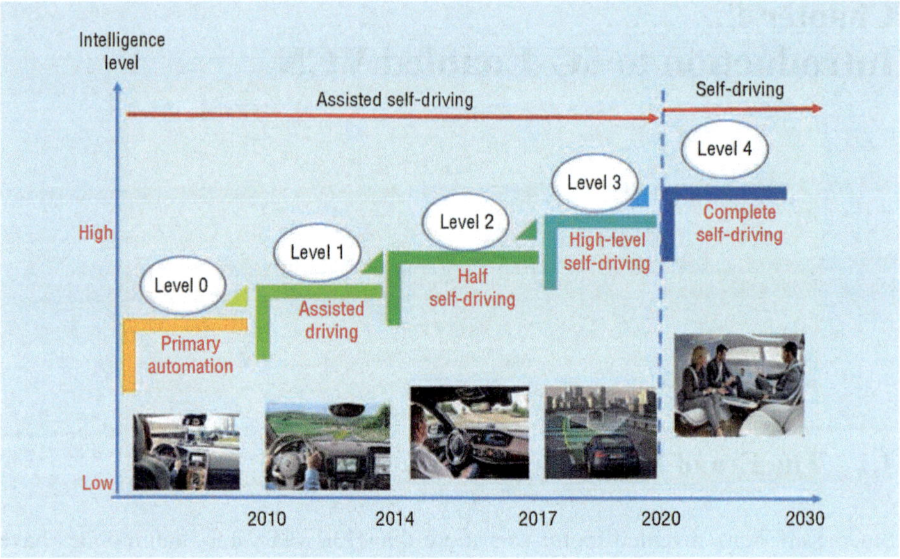

Fig. 1.1 Five developing levels of intelligent vehicles: primary automation, assisted driving, half self-driving, high-level self-driving, and complete self-driving

From the beginning of intelligent transportation system (ITS) as a research field in the mid-1980s, IVs have been one of the most significant ITS applications. The development of IVs can be divided into two stages: the initial stage for assisted driving, and the ultimate stage for complete self-driving instead of human driving. The National Highway Traffic Safety Administration (NHTSA) divides IVs into five levels as illustrated in Fig. 1.1.

The ultimate objective of the IV development is self-driving/autonomous vehicles. Currently, this is generally considered a futuristic concept that remains a far reach from actual deployment [4, 5]. Various research projects have advanced the enabling technologies in environmental perception and vehicle control and have produced experimental implementations to show how automation technologies could be applied to road vehicles. These have led to major demonstrations in Europe, North America, Japan, and China. Academic research has been ongoing as well, largely out of sight of the general public [6, 7].

Recently, several companies, including Google, Tesla Motors, and Baidu, have devoted heavy effort and resources to develop self-driving cars. These self-driving cars have attracted an unprecedented level of media interest, while raising speculation about the impacts and implications of automated driving on societal matters such as road safety, privacy, traffic flow, energy and environmental issues, land use, economics of the vehicle industry, and cybersecurity. Self-driving is the development trend of IVs. From a technical viewpoint, there are two basic self-driving architectures. The first is based on the vehicle platform. The on-vehicle sensors perform environment perception and data fusion, then make decisions and control the vehicle via the vehicle execution unit. The second is based on vehicular ad-hoc networks (VANETs). The vehicle receives environmental data and roadside information via the VANETs.

At present, the first architecture is the mainstream approach, on which most researchers and developers are focusing. With this approach, the environment information surrounding the vehicle is acquired by sensors onboard the vehicle. On the basis of this information, the vehicle will independently accomplish the automatic-driving control, including environment sensing, central decision making, and mechanical control. To obtain reliable and comprehensive environment information, the vehicle is often equipped with expensive multi-beam LiDARs (light detection and ranging sensors), microwave Radars, and high-resolution cameras. At the same time, the vehicle must be equipped with complex and costly processing and control units to ensure the safety and reliability of automatic driving. These not only significantly increase the vehicle cost, but also hinder the development and adoption of affordable IVs. Additionally, environment sensing and automatic driving control based on the sole intelligence of an individual vehicle would inevitably lead to limitations and safety concerns.

In a nutshell, IVs that depend on individual vehicle sensing and control face the challenges of high cost and limited processing capability. A satisfying yet affordable solution largely depends on cost reduction of the sensing, processing, and control components. Therefore, a more active architecture is urgently demanded to address these limitations for IVs with improved safety and reliability, yet shortened time to market.

1.2 5G-Enabled Vehicular Communications and Networking (5G-VCN)

As one of the key supportive technologies for the next generation ITS and IVs, vehicular communications and networking (VCN) is receiving increasing interest since the end of last century. In order to facilitate various applications including vehicle safety, transportation efficiency, and entertainment, VCN needs to achieve low latency and high reliability communications. The vehicular ad-hoc network (VANET) based on the mobile ad hoc architecture first attracted lots of research and standardization efforts. To effectively support VANETs, the dedicated short-range communications (DSRC) standard based on IEEE 802.11p [8] and IEEE 1609.x [9–13] has gained the support and promotion of the federal government in the United States. Systems based on DSRC and alternative systems have been developed and standardized in the European Union, China, and Japan, etc. However, since VANET requires significant infrastructure investment up front, it is still in the field trial stage and has not been widely developed yet. Moreover, the achievable rate and the network configuration of DSRC-based VANET cannot catch up with the fast and ever-increasing communication requirements of vehicular applications,

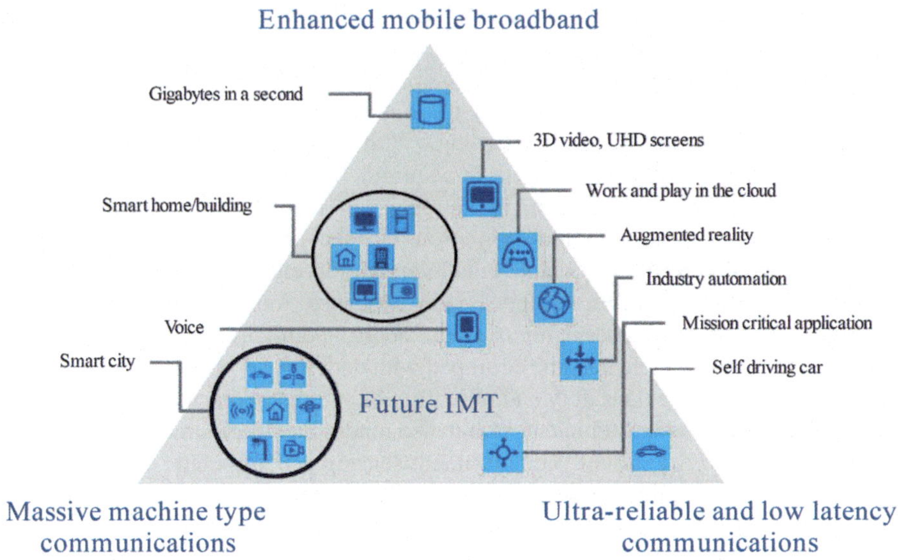

Fig. 1.2 5G application scenarios. (Source from ITU-R M.2083-0)

especially for near-future autonomous vehicles. In addition, IEEE 802.11p has shown poor scalability and lacks guaranteed service delivery in large-scale network deployments.

Different from VANET, cellular network boasts mature development, well-established business model, and comprehensive standardization progress. Since 1983 when the first generation (1G) cellular system had been standardized and developed, the cellular system typically rolls out a new generation every decade. As of today, the 4G long-term evolution (LTE) system has been widely developed and completely commercialized. Supporting vehicular communications with mature and widely deployed LTE, that is, LTE for vehicle (LTE-V) [14], is becoming a rising star for VCN. Currently, the standardization and commercialization of LTE-V are both under active development. As the next generation of the cellular network, the 5G, is coming forward. Different from all previous generations, 5G for the first time lists the VCN as one of the representative application scenarios as shown in Fig. 1.2. In order to support additional potential applications related to future vehicles, such as intelligent and autonomous vehicles, 5G-enabled vehicular communications and networking (5G-VCN) is calling for higher reliability and lower latency transmission of huge amount of data. As a result, the research and design of 5G-VCN are very challenging, and are receiving significant attention.

1.2.1 5G-VCN: Key Features

As a new-generation mobile communication technology, 5G is currently undergoing active development. The envisioned 5G applications centered at ultra-high data rate support with ultra-low latency and ultra-reliability provide a promising enabler for self-driving IVs. More importantly, 5G's unique features, such as proximity service (ProSe), data-control-separated software-defined network (SDN), flexible network architecture and topology, cloud/fog computing and processing, and application-oriented design, render it not only a vital supporting technology for IVs but also an integrated part of the IV system.

An important feature of 5G-enabled communications is the proximity service (ProSe) evolved from device-to-device (D2D) communications [16]. Currently, the aim of ProSe is to provide awareness by discovering devices and services using relevant locality information. In the context of the 5G-VCN, ProSe is particularly significant for many spontaneous interactions or communication opportunities within a certain proximity [15]. The key enablers for ProSe-based applications are the location information and the communication trends in social networks. Unlike traditional location discovery over the network, ProSe provides ad hoc location discovery and communication opportunities (e.g., moving vehicles on roads). More importantly, with the ability to discover and communicate without an infrastructure, ProSe can be used as a communication platform in public safety scenarios. Moreover, such ad hoc discovery and communication among mobile users provide a means for high data rate transmissions (i.e., by avoiding latency due to traversing through the core) and high efficiency in resource utilization (i.e., by avoiding transmissions passing through the core network), thereby reducing congestion in the core network.

The SDN in 5G network architectures features data-control separation [17]. This can cope with the network-control issues by imposing a centralized control plane of individual network devices at an external entity. With the development of SDN for IVs, the network latency could be improved with specific self-driving operations deployed on the centralized control plane. SDN renders the remote control of IVs by 5G-VCN a practical possibility.

In addition, the coexistence of a centralized cloud network architecture and distributed fog/edge networks in 5G makes it possible for 5G-VCN to realize IV data storage and processing at three levels, i.e., cloud, fog, and onboard [18, 19]. Depending on the data characteristics and latency requirements of different IV services, 5G-VCN can arrange their storage and processing in a flexible manner. Based on the cloud and fog platforms, 5G-VCN can better exploit the availability of multi-IV data to facilitate various learning functions more effectively and efficiently, and thereby lead to self-driving behaviors that better imitate or even surpass human driving.

Another key feature of 5G is the hierarchical coexistence of heterogeneous networks [20]. Here, the heterogeneity refers not only to cells of various sizes but also to different protocols and standards. As a result, 5G-VCN can better support the stable high-speed network connectivity of IVs. For example, the control messages of IVs could be maintained by macrocell base stations with wide-area coverage to ensure stable connectivity under high mobility, whereas the exchange of massive data can be enabled by micro- and femto-cells via advanced data dissemination and D2D technology, such that low latency and high reliability can be guaranteed.

Finally, 5G-VCN can send the 3D high-resolution map and real-time traffic data to IVs to assist high-precision localization and the corresponding real-time route planning. This should complement the limitations of onboard sensors, and thus improve both the reliability and robustness of IVs. At the same time, this could also alleviate the dependency of IVs on expensive onboard sensing equipment and thereby lower the vehicle cost, and in turn shorten the time to market.

1.2.2 5G-VCN: Challenges

Although 5G-VCN has been regarded as a promising solution for future IV applications, it still faces quite many challenges.

1.2.2.1 Ultra-high Rate and Ultra-low Latency Vehicular Communications

For autonomous vehicles moving in high-speed self-driving mode, fast and real-time exchange of dynamic information must be facilitated among vehicles and between vehicles and the fog and cloud. Such information includes both small data (such as speed, location, and direction of neighboring vehicles) and big data (such as video of surrounding environments and/or 3D high-resolution maps). Ultra-low latency typically requires a time delay in the scale of microseconds and a data exchange rate of 10 times per second. Although many techniques in 5G systems (e.g., Massive MIMO, mm-Wave communications, D2D communications, and so on) have been proposed and tested to achieve ultra-high data rate and ultra-low latency communications, the investigations or experiments are mostly based on static or low-mobility scenarios. The channel situations and network architecture in vehicular networks are quite different from traditional cellular networks. Due to high mobility in vehicular networks, vehicular communications face severe inter-carrier interference (ICI) and rapidly time-varying fading channels, which would markedly degrade the performance of current 5G communication techniques. In addition, due to the high vehicle mobility, heterogeneous and frequently changing topology, and versatile quality-of-service (QoS) requirements, the resource accessing and data dissemination designs in VCN all pose significant challenges [21].

1.2.2.2 Wireless System Architecture for Self-Driving

The service characteristics and user behavior patterns for the automobile industry differ from conventional communications. We need to carefully analyze and sort out the specific requirements of various vehicles and their services and applications, and then design the 5G-VCN system architecture. To adapt to the real-time communication requirements, and to realize the information exchange and coordination control among onboard equipment, roadside units, and the onboard communication platform, a fully distributed system architecture centered at the vehicle-road coordination network must be comprehensively designed. Such an architecture should be an open, safe, and highly efficient network that features data broadcast, end-to-end data flow storage, and self-organization, transmission, and control of high-mobility network nodes. In the meantime, scalability of such a network is also critical to ensure future integration with the smart grid with electrical vehicles being the inevitable trend [22–24].

1.2.2.3 Environment Sensing Technology

5G-VCN is expected to help provide beyond visual range (BVR) environment information for IVs. In other words, 5G essentially serves as a virtual sensor or tele-sensor for IVs. On the other hand, there are various onboard sensors, including multi-beam LiDAR, millimeter radar, and video camera. The optimized fusion of these local and remote heterogeneous sensors with varying levels of resolution and latency is yet another key task for 5G-VCN. Once properly solved, such an optimized configuration could significantly enhance the environmental awareness of each IV, while transferring the high yet redundant per-IV equipment cost to the shared roadside facilities via collection, fusion, and sharing. We believe that this is pivotal in bringing safe yet affordable self-driving IVs into reality.

1.3 Organization of the Monograph

As illustrated in Fig. 1.3, the rest of this monograph is organized as follows.

1. In Chap. 2, the vehicular channel characteristics and modeling are discussed and a new generic geometry-based stochastic modeling approach is provided, which serve as the fundamental for 5G-VCN technology development.
2. In Chap. 3, from the wireless-vehicle combination perspective, advanced PHY techniques suitable for 5G-VCN are introduced and discussed in detail, including ICI cancellation, index modulated (IM-)OFDM, differential spatial modulation (DSM), energy harvesting (EH)-based vehicular communications, and some next-leap directions.

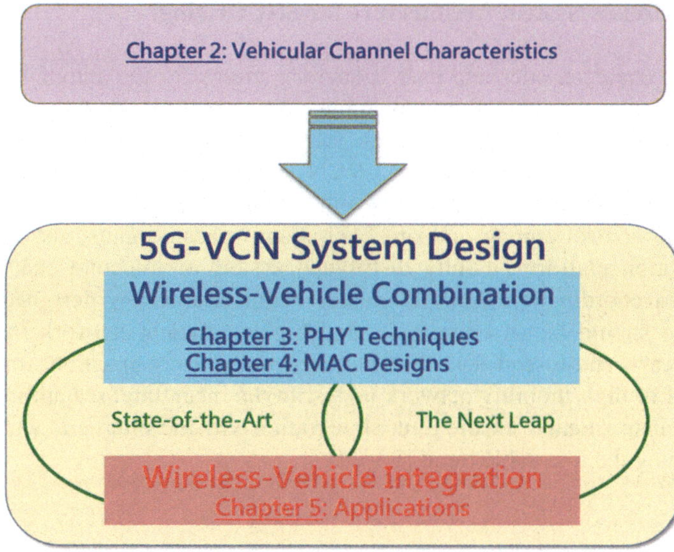

Fig. 1.3 Organization of this monograph

3. In Chap. 4, following the wireless-vehicle combination perspective, we elaborate on effective MAC designs in both distributed and centralized manners targeting at 5G-VCN, including distributed congestion control, centralized resource sharing and scheduling, and centralized data dissemination.
4. In Chap. 5, from the wireless-vehicle integration perspective, we explore and discuss some important 5G-VCN-based applications, including electric vehicles, distributed data storage, and physical layer security. In addition, autonomous driving is also discussed as the next leap for 5G-VCN-based IV applications.

References

1. X. Cheng, C. Chen, W. Zhang and Y. Yang, "5G-Enabled Cooperative Intelligent Vehicular (5GenCIV) Framework: When Benz Meets Marconi," *IEEE Intelligent Systems*, vol. 32, no. 3, pp. 53–59, May-June 2017.
2. A. Broggi, A. Zelinsky, M. Parent, and C. E. Thorpe, Chapter 51: Intelligent Vehicles, Springer Handbook of Robotics, pp. 1175–1198, Jan. 2008.
3. D. Pomerleau and T. Jochem, "Rapidly adapting machine vision for automated vehicle steering," *IEEE Expert*, 11(2) 19–27 (1996), Special Issue on Intelligent System and their Applications.
4. J. Petit and S.E. Shladover, "Potential Cyberattacks on Automated Vehicles," *IEEE Trans. Intelligent Transportation Systems*, vol. 16, no. 2, 2015, pp. 546–556.
5. E. Ohn-Bar and M.M. Trivedi, "Looking at Humans in the Age of Self-Driving and Highly Automated Vehicles," *IEEE Trans. Intelligent Vehicles*, vol. 1, no. 1, 2016, pp. 90–104.

6. K. Jo *et al.*, "Development of Autonomous Car Part I: Distributed System Architecture and Development Process," *IEEE Trans. Industrial Electronics*, vol. 61, no. 12, 2014, pp. 7131–7140.
7. K. Jo *et al.*, "Development of Autonomous Car Part II: A Case Study on the Implementation of an Autonomous Driving System Based on Distributed Architecture," *IEEE Trans. Industrial Electronics*, vol. 62, no. 8, 2015, pp. 5119–5132.
8. Wireless LAN Medium Access Control (MAC) and Physical Layer (PHY) Specifications Amendment 6: Wireless Access in Vehicular Environments, IEEE Standard 802.11p, 2010.
9. IEEE Guide for Wireless Access in Vehicular Environments (WAVE) – Architecture, *IEEE Std 1609.0-2013*, pp. 1–78, Mar. 2014.
10. IEEE Standard for Wireless Access in Vehicular Environments – Security Services for Applications and Management Messages, *IEEE Std 1609.2-2016 (Revision of IEEE Std 1609.2-2013)*, pp. 1–240, Mar. 2016.
11. IEEE Standard for Wireless Access in Vehicular Environments (WAVE) – Networking Services, *IEEE Std 1609.3-2016 (Revision of IEEE Std 1609.3-2010)*, pp. 1–160, Apr. 2016.
12. IEEE Standard for Wireless Access in Vehicular Environments (WAVE) – Multi-Channel Operation, *IEEE Std 1609.4-2010 (Revision of IEEE Std 1609.4-2006)*, pp. 1–89, 7 Feb. 2011
13. IEEE Standard for Wireless Access in Vehicular Environments (WAVE) – Identifier Allocations, *IEEE Std 1609.12-2012*, pp. 1–20, Sept. 2012.
14. S. Chen, J. Hu, Y. Shi, and L. Zhao, "LTE-V: A TD-LTE-Based V2X Solution for Future Vehicular Network," *IEEE Internet of Things Journal*, vol. 3, no. 6, pp. 997–1005, Dec. 2016.
15. S. A. A. Shah, E. Ahmed, M. Imran and S. Zeadally, "5G for Vehicular Communications," *IEEE Communications Magazine*, vol. 56, no. 1, pp. 111–117, Jan. 2018.
16. X. Lin *et al.*, "An Overview of 3GPP Device-to-Device Proximity Services," *IEEE Commun. Mag.*, vol. 52, no. 4, Apr. 2014, pp. 40–48.
17. D. Kreutz *et al.*, "Software-Defined Networking: A Comprehensive Survey," *Proc. IEEE*, vol. 103, no. 1, 2015, pp. 14–76.
18. Q. Han, S. Liang, and H. Zhang, "Mobile Cloud Sensing, Big Data, and 5G Networks Make an Intelligent and Smart World," *IEEE Network*, vol. 29, no. 2, 2015, pp. 40–45.
19. S. Huang *et al.*, "Architecture Harmonization between Cloud Radio Access Networks and Fog Networks," *IEEE Access*, vol. 3, Dec. 2015, pp. 3019–3034.
20. F. Gabry, V. Bioglio, and I. Land, "On Energy-Efficient Edge Caching in Heterogeneous Networks," *IEEE J. Selected Areas in Comm.*, vol. 34, no. 12, 2016, pp. 3288–3298.
21. X. Cheng, R. Zhang, L. Yang, "Wireless Towards the Era of Intelligent Vehicles," Submitted to Internet of Things Journal.
22. X. Cheng *et al.*, "Electrified Vehicles and the Smart Grid: The ITS Perspective," *IEEE Trans. Intelligent Transportation Systems*, vol. 15, no. 4, 2014, pp. 1388–1404.
23. X. Cheng, R. Zhang, and L. Yang, "Consumer-Centered Energy System (CCES) for Electric Vehicles and Smart Grid," *IEEE Intelligent Systems*, vol. 31, no. 3, 2016, pp. 97–101.
24. R. Zhang, X. Cheng, and L. Yang, "Energy Management Framework for Electric Vehicles in the Smart Grid: A Three-Party Game," *IEEE Comm. Magazine*, vol. 54, no. 12, 2016, pp. 93–101.

Chapter 2
Vehicular Channel Characteristics and Modeling

Nowadays, intelligent transportation systems and vehicular networks are developing rapidly with ever-increasing researches on the vehicular communication channels. Vehicle-to-everything (V2X) communications is a key technology for future vehicular networks and applications, which enables vehicles to obtain a broad range of traffic information such as real-time road conditions, road information, pedestrian information, etc., and thus improve driving safety, reduce congestion, improve traffic efficiency, and provide in-vehicle entertainment. It is well-known that the successful design of any wireless communication system requires a proper description of the propagation environment and a corresponding realistic yet easy-to-use channel model. Hence, in this chapter, we will first review recent advances in vehicular channel measurements and modeling, and then provide a new generic wideband geometry-based stochastic modeling. In addition, by taking the new features due to 5G into consideration, the challenges and open issues for 5G vehicular channel measurements and modeling are discussed.

2.1 Recent Advances in Channel Measurements and Modeling

In this section, we will provide an overview of recent advances in channel measurements and modeling for vehicular communications.

2.1.1 Channel Measurements

So far, some research groups have conducted V2V measurement campaigns [1–9] and worked on developing generic V2V channel models for V2V channel characterization. Next, some recent typical measurement campaigns will be reviewed

X. Cheng et al., *5G-Enabled Vehicular Communications and Networking*,
Wireless Networks, https://doi.org/10.1007/978-3-030-02176-4_2

according to carrier frequencies, frequency selectivity, antennas, environments, Tx/Rx directions of motion, and channel statistics.

2.1.1.1 Carrier Frequencies

Before the IEEE 802.11p standard was proposed, some measurement campaigns were conducted at carrier frequencies outside the 5.9 GHz dedicated short range communication (DSRC) band. In [4], V2V measurements were carried out at 2.4 GHz, i.e., the IEEE 802.11b/g band. Some measurements were done around the IEEE 802.11a frequency band, e.g., at 5 GHz in [5] and at 5.2 GHz in [6]. Measurements at 5.9 GHz were presented in [7] and [8] for narrowband and wideband V2V channels, respectively. The aforementioned measurements have shown that propagation phenomenon in similar environments with different frequencies can vary significantly. Therefore, more measurement campaigns are expected to be conducted at 5.9 GHz for the better design of safety applications for V2V systems following the IEEE 802.11p standard. On the other hand, for improved design of non-safety applications for V2V systems, measurement campaigns performed at other frequency bands, e.g., 2.4 or 5.2 GHz, are still required.

2.1.1.2 Frequency Selectivity and Antennas

In the USA, the Federal Communications Commission has allocated 75 MHz of licensed spectrum for DSRC, including seven channels, each with approximately 10 MHz instantaneous bandwidth. Such V2V channels are nearly always frequency-selective (or wideband) channels. Channel characterization based on narrowband measurement results [7] is not sufficient for such V2V DSRC applications. Wideband measurement campaigns [4–6] are therefore essential for understanding the frequency-selectivity features of V2V channels and designing high-performance V2V systems. Most V2V measurement campaigns so far have focused on single-antenna applications, resulting in single-input single-output (SISO) systems [7, 8]. MIMO systems, with multiple antennas at both ends, are very promising candidates for future communication systems and are gaining more importance in IEEE 802.11 standards. Moreover, MIMO technology becomes more attractive for V2V systems since multiple antenna elements can be easily placed on large vehicle surfaces. Hence, more MIMO V2V wideband measurement campaigns are needed for future V2V system developments.

2.1.1.3 Environments and Tx/Rx Directions of Motion

Similar to conventional F2M cellular systems, V2V scenarios can be classified as large spatial scale (LSS), moderate spatial scale (MSS), and small spatial scale (SSS) according to the Tx-Rx distance. For LSS scenarios or MSS scenarios, where the Tx-Rx distance is normally larger than 1 km or ranges from 300 m to 1 km, V2V systems are mainly used for broadcasting or geocasting, i.e., geographic broadcasting. For SSS scenarios, where the Tx-Rx distance is usually smaller than 300 m, V2V systems can be applied to broadcasting, geocasting, or unicasting. Since most V2V applications fall into MSS or SSS scenarios, these two scenarios are currently receiving more and more attention with several current measurement campaigns taking place [4–8]. However, there are still few applications that need communications between two vehicles separated by large distances, e.g., larger than 1 km. For such LSS V2V applications, one example is V2V decentralized environmental notification, which means that vehicles or drivers in a certain area share information with each other about observed events or roadway features. These applications have not gained much attention and thus no measurement results are available that explore V2V channels for LSS scenarios. Due to the unique feature of V2V environments, the vehicular traffic density (VTD) also significantly affects the channel statistics, especially for MSS and SSS scenarios. In general, the smaller Tx-Rx distance, the larger impact of the VTD. Note that V2V channels usually exhibit non-isotropic scattering except in cases of high VTD. Directions of motion of the Tx and Rx also affect channel statistics, e.g., Doppler effects. Many measurement campaigns [4, 5] have focused on studying channel characteristics when the Tx and Rx are moving in the same direction. Few V2V measurement campaigns [6, 8], have investigated channel characteristics when the Tx and Rx are moving in opposite directions.

2.1.1.4 Channel Statistics

Knowledge of channel statistics is essential for the analysis and design of a communication system. Many different V2V channel statistics have been studied in recent measurement campaigns. Here, only amplitude distribution and Doppler power spectral density (DPSD) are concentrated on. Analysis of amplitude distributions has been reported in [5, 7, 8]. In [8], the authors modeled the amplitude probability density function (PDF) of the received signal as either Rayleigh or Ricean. In [7], it was observed that the received amplitude distribution in a dedicated V2V system with a carrier frequency of 5.9 GHz gradually transits from near-Ricean to Rayleigh as the vehicle separation increases. When the line-of-sight (LoS) component is intermittently lost at large distances, the channel fading can become more severe than Rayleigh. A similar conclusion has been drawn in [5], where the amplitude PDF is modeled as Weibull distribution and this worse than Rayleigh fading is called severe fading. The reason behind the severe fading is the rapid transitions of multipath components induced by high speed and low height of the Tx/Rx and fast moving scatterers.

The DPSD has been investigated in [4, 6–8]. It was demonstrated that DPSDs can vary significantly with different time delays in a wideband V2V channel. In [7], the authors analyzed the Doppler spread and coherence time of narrowband V2V channels and presented their dependence on both velocity and vehicle separation.

Recently, the space-DPSD, which is the Fourier transform of the space-time correlation function in terms of time, was investigated in [6]. It is worth noting that the DPSD for V2V channels can be significantly different from the traditional U-shaped DPSD for F2M channels.

2.1.2 *Recent Channel Models*

In terms of the modeling approach, these models can be categorized into geometry-based deterministic models (GBDMs) and stochastic models, and the latter can be further classified as non-geometry-based stochastic models (NGSMs) and geometry-based stochastic models (GBSMs). Here, these various channel models will be introduced.

2.1.2.1 GBDMs

GBDMs characterize V2V physical channel parameters in a completely deterministic manner. Maurer et al. [10] proposed a GBDM based on the ray-tracing method for V2V channels. This method is intended to reproduce the actual physical radio propagation process for a given environment. As shown in Fig. 2.1, assume that the representation of the real environment mainly consists of two main parts: the modeling of dynamic road traffic (such as moving cars, trucks, trucks, etc.) and modeling of roadside environments (buildings, parked cars, road signs, trees, etc.) Then, an accurate model of the wave propagation in the real environment

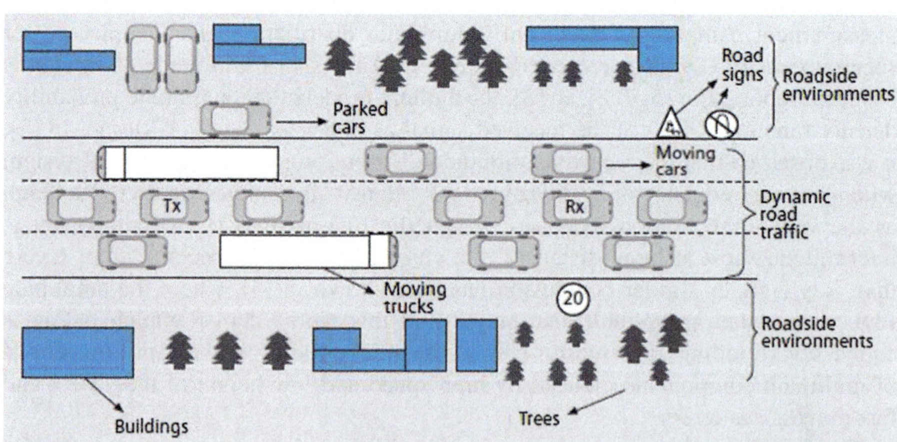

Fig. 2.1 A typical V2V communication environment and corresponding geometric description of GBDMs

is achieved by generating possible paths (rays) from the Tx to Rx according to geometric considerations and the rules of geometrical optics. However, GBDM requires a detailed and time-consuming description of the site-specific propagation environment and therefore cannot be easily generalized to a wide range of scenarios.

2.1.2.2 NGSMs

NGSMs do not assume any geometry, but rather determine the physical parameters of a V2V channel in a completely stochastic manner. Acosta-Marum and Brunelli [8] proposed a SISO NGSM as the origin of the V2V channel model. Based on the tapped delay line structure, the model consists of L taps, with tap amplitude PDF being either Rayleigh or Rician, so it is capable of reflecting the channel statistical characteristics of each tap. In addition, each tap contains N unresolvable sub-paths with different types of Doppler spectrum: flat, round, classic-3-dB, and classic-6-dB shapes. This allows the synthesis of almost arbitrary Doppler spectra for each tap. However, the NGSM is still based on the wide sense stationary uncorrelated scattering (WSSUS) assumption and is not used to investigate the impact of the VTD on channel statistics.

Recently, a new SISO NGSM was proposed by Sen and Matolak [5]. The model considered the non-stationary characteristics of the channel by modeling the persistence of multi-path components through Markov chains. The impact of VTD on channel statistics was also investigated. The complex impulse response of the SISO V2V channel in the study was derived from an additional term proposed by Acosta-Marum and Brunelli: a continuous process that takes into account the finite-life resolvable path. The NGSM can easily capture the effect of sudden disappearance of strong multi-paths, which is usually caused by rapid blockage or obstruction by another vehicle or some other obstacle. However, this model did not take into account the movement of scatterers in different delay bins (resolvable paths), and therefore the transitional probabilities of the Markov model for continuous processes may be accurate. This may reduce the ability of the NGSM to accurately capture the non-stationarity of real V2V channels and this question therefore deserves more investigation.

2.1.2.3 GBSMs

GBSMs are derived from the predefined stochastic distribution of effective scatterers according to the basic laws of wave propagation. Such a model can be easily adapted to different scenes by changing the shape of the scattering region or the PDF of the position of scatterers. GBSMs can be further classified into regular-shape GBSMs (RS-GBSMs) or irregularly-shaped GBSMs (IS-GBSMs) according to whether effective scatterers are placed on regular shapes (one/two rings, ellipse, etc.) or irregular shapes.

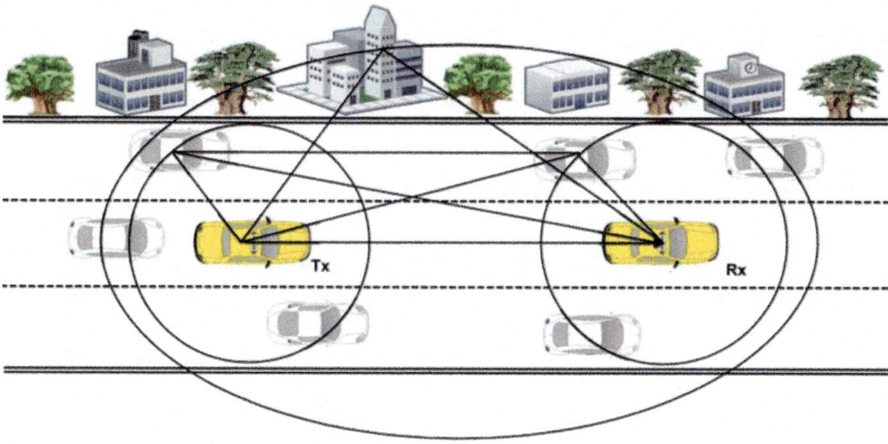

Fig. 2.2 A typical V2V communication environment described using GBSM

IS-GBSMs [1, 11] place effective scatterers with specified characteristics at random locations with a specified statistical distribution. The contribution of the effective scatterers is determined by a simplified ray tracing method and the final signal is summed up to obtain he complex impulse response. Using the ray tracing method, the IS-GBSM proposed in [11] can more easily handle the non-stationarity of the V2V channel by specifying the motion of Tx, Rx and moving scatterers.

For RS-GBSMs, Akki and Haber [12] first proposed a two-dimensional (2D) RS-GBSM for the narrowband isotropic SISO V2V Rayleigh fading channel. A two-ring GBSM for non-isotropic MIMO V2V Ricean channels was proposed in [13, 14]. In addition, a RS-GBSM was proposed which is the combination of a ellipse and two-ring model for non-isotropic V2V-MIMO channels in [15, 16]. The paper [17] proposed a general three-dimensional (3D) two-cylinder model for narrowband V2V channels. The envelope level crossing rate (LCR) and average fade duration (AFD) of the V2V channel were investigated in [18–21]. Sen and Matolak [5] proposed some research on VTD, which can prove VTD to V2V channel. Channel statistics have a certain impact. In order to further study the model of wideband V2V channels, a two-concentric-cylinder GBSM for wideband channels was proposed in [22, 23]. Figure 2.2 shows a typical V2V communication environment including moving and static scatterers.

2.2 New Generic Wideband Geometry-Based Stochastic Modeling

Nowadays, with the rapid development of intelligent transportation system and vehicle self-organizing network, the research on vehicle communication channel is increasing day by day. Because of the characteristics of vehicle traveling at high speed and limited moving area, the V2V communication system has a significant difference from the traditional cellular system. The biggest difference between the V2V communication systems and the cellular network in the transmission environment is that both the transmitter (Tx) and receiver (Rx) are in motion and equipped with low elevation antennas, and there are a large number of scatterers near the Tx and Rx, which also may be moving. V2V channel usually exhibits the characteristics of fast change, non-isotropic, as well as complex and time-varying Doppler. Therefore, it is important to develop a practical and easy-to-use V2V MIMO channel model to understand the unique V2V channel characteristics and design the vehicle communication system accordingly. Since V2V communication scenarios are in general time variant due to the movement of the Tx and Rx, the GBSM method is more suitable as this method directly deals with propagation environments. Therefore, the use of the GBSM method for modeling V2V propagation channels has attracted more and more attention.

In [24], a 2D two-ring and ellipse model for V2V narrowband channel has been extended to wideband channel by using confocal multiple ellipses model to represent the tapped delay line (TDL) structure, and the effect of DPSD on the model was also investigated. However, it did not derive the expression of the corresponding statistical properties of space-time correlation function (STCF), LCR and AFD. Note that these aforementioned GBSMs [22–24] are reference model since these models assume an infinite number of effective scatterers, i.e., has an infinite complexity, and thus cannot be directly implemented in practice. Therefore, accurate simulation models play an important role in the practical simulation and performance evaluation of any wireless communication systems. It is worth noting that the 2D GBSM modeling and investigation of V2V-MIMO wideband channels are surprisingly missing in the current literature.

To fill up the aforementioned gaps of V2V-MIMO GBSMs, a new generic 2D GBSM for V2V-MIMO wideband channels was proposed in paper [25]. The proposed 2D wideband GBSM mainly investigates on the basis of a two-ring and a multiple confocal ellipse model for V2V MIMO wideband channel proposed in [24], which combines LoS components, single-bounced (SB) and double-bounced (DB) components.

2.2.1 A Wideband V2V-MIMO Channel Reference Model

Assuming that both the Tx and the Rx that equipped with n_T transmit and n_R receive low elevation antenna elements are moving, where $1 \leq p \leq p' \leq n_T$ and $1 \leq q \leq q' \leq n_R$, respectively. It is assumed that the scatterers are distributed over the two-ring model and the confocal multi-ellipsoidal model randomly. The two-ring model is used to represent moving scatterers, such as the moving vehicles, and the multiple confocal ellipse models are used to represent the static scatterers, such as the static roadside environment. For a two-ring model, it is assumed that

$N_{1,1}$ effective scatterers are distributed around the ring of radius R_T at the Tx, and the $n_{1,1}$th ($n_{1,1} = 1, 2, \ldots, N_{1,1}$) effective scatterer is defined as $s^{(n_{1,1})}$. Similarly, suppose there are $N_{1,2}$ effective scatterers are distributed around the ring of radius R_R at Rx, and the $n_{1,2}$th ($n_{1,2} = 1, 2, \ldots, N_{1,2}$) effective scatterer is denoted by $s^{(n_{1,2})}$. In addition, for the elliptical model, the multiple confocal ellipses model is used here to represent the TDL structure, where their focal points are located at Tx and Rx. It can be seen that the distance between the transmitter and receiver can be expressed as $D = 2f$, which the parameter f represents the half length of the two foci connection of the ellipse. Assuming that the lth ellipse's semi-major axis (i.e., the lth tap) is a_l and there are $N_{l,3}$ effective scatterers distributed over it, ($l = 1, 2, \ldots, L$), where L is the total number of confocal ellipses. Note that the $n_{l,3}$th ($n_{l,3} = 1, 2, \ldots, N_{l,3}$) effective scatterer can be described as $s^{(n_{l,3})}$.

Figure 2.3 shows the basic 2 × 2 V2V-MIMO wideband channel model ($n_T = n_R = 2$), that is, the transmitter and receiver set up two uniform linear antenna components. In addition, the velocities of the Tx and Rx are denoted as v_T and v_R, and the movement direction angles are γ_T and γ_R, respectively. By considering

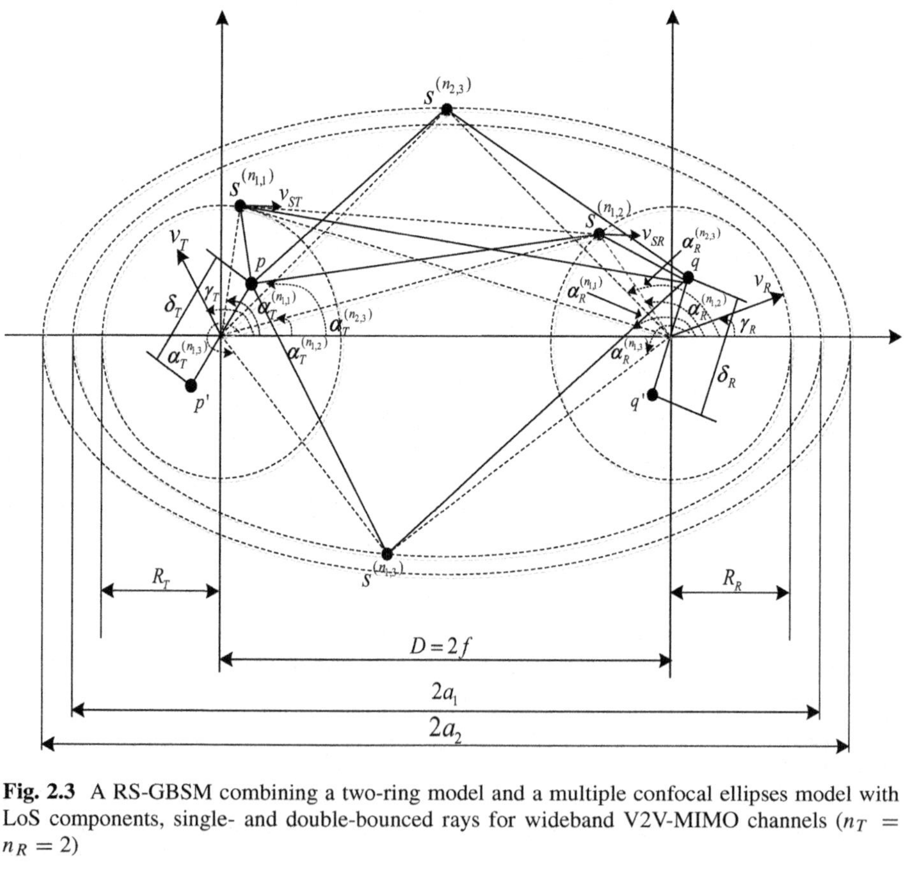

Fig. 2.3 A RS-GBSM combining a two-ring model and a multiple confocal ellipses model with LoS components, single- and double-bounced rays for wideband V2V-MIMO channels ($n_T = n_R = 2$)

the real V2V communication scenario, it is assumed that the moving scatterers distributed around Tx and Rx move at speed v_{ST} and v_{SR}, respectively, and their movement directions are along the x-axis. The angle of departure (AoD) $\alpha_T^{(n_{l,i})}$ $(i = 1, 2, 3)$ characterizes the relative position of scatterer $s^{(n_{l,i})}$ to Tx. Similarly, the angle of arrival (AoA) $\alpha_R^{(n_{l,i})}$ characterizes the relative position of scatterer $s^{(n_{l,i})}$ to Rx. The AoD and AoA of the LoS path are denoted by α_T^{LoS} and α_R^{LoS}.

Because the different rays have different contributions in a real V2V communication environments, different taps of our model to express the V2V channel statistics need to be designed. For the first tap, the corresponding single- and double-bounced components can be considered as produced by the scatterers located on the first ellipse or one of the two rings, and the scatterers located on both two-rings, respectively. In addition, considering other taps, only the scatterers on the confocal ellipses can generate the single-bounced rays, while the double-bounced rays come from the corresponding ellipse and one of the two rings.

Here, an $n_T \times n_R$ matrix $H(t, \tau) = [h_{pq}(t, \tau)]_{n_T \times n_R}$ is used to describe the V2V-MIMO wideband channel. The channel impulse response between the pth Tx and the qth Rx antenna elements can be expressed as $h_{pq}(t, \tau) = \sum_{l=1}^{L} c_l h_{l,pq}(t)\delta(\tau - \tau_l)$, where c_l represents the gain of the lth tap, and the complex time-variant tap coefficient and the discrete propagation delay of the lth tap are denoted as $h_{l,pq}(t)$ and τ_l, respectively. For the first tap, the complex tap coefficient of the $T_p - R_q$ link can be expressed as

$$h_{1,pq}(t) = h_{1,pq}^{LoS}(t) + \sum_{i_1=1}^{I_1} h_{1,pq}^{SB_{i_1}}(t) + h_{1,pq}^{DB}(t) \tag{2.1}$$

with

$$h_{1,pq}^{LoS}(t) = \sqrt{\frac{K\Omega_{pq}}{K+1}} e^{-j2\pi f_c \tau_{pq}} e^{j2\pi t[f_{Tm}\cos(\alpha_T^{LoS}-\gamma_T)+f_{Rm}\cos(\alpha_R^{LoS}-\gamma_R)]} \tag{2.2a}$$

$$h_{1,pq}^{SB_{i_1}}(t) = \sqrt{\frac{\eta_{SB_{1,i_1}}\Omega_{pq}}{K+1}} \lim_{N_{1,i_1}\to\infty} \frac{1}{\sqrt{N_{1,i_1}}} \sum_{n_{1,i_1}=1}^{N_{1,i_1}} e^{j\phi^{(n_{1,i_1})}} e^{-2\pi f_c \tau_{pq,n_{1,i_1}}} e^{j2\pi t[f_{Tm,i_1}\cos(\alpha_T^{(n_{1,i_1})}-\gamma_T)+f_{Rm,i_1}\cos(\alpha_R^{(n_{1,i_1})}-\gamma_R)]} \tag{2.2b}$$

$$h_{1,pq}^{DB}(t) = \sqrt{\frac{\eta_{DB}\Omega_{pq}}{K+1}} \lim_{N_{1,1},N_{1,2}\to\infty} \frac{1}{\sqrt{N_{1,1}N_{1,2}}} \sum_{n_{1,1},n_{1,2}=1}^{N_{1,1},N_{1,2}} e^{j\phi^{(n_{1,1},n_{1,2})}} e^{-2\pi f_c \tau_{pq,n_{1,1},n_{1,2}}} e^{j2\pi t[f_{Tm,1}\cos(\alpha_T^{(n_{1,1})}-\gamma_T)+f_{Rm,2}\cos(\alpha_R^{(n_{1,2})}-\gamma_R)]} \tag{2.2c}$$

where $I_1 = 3$, the symbols Ω_{pq} and K denote the total power of the T_p-R_q link and the Ricean factor, respectively. The scattering-caused phases $\phi^{(n_{1,i_1})}$ and $\phi^{(n_{1,1},n_{1,2})}$ are random variables with uniform distributions over $[-\pi, \pi)$. $\tau_{pq} = \varepsilon_{pq}/c$, $\tau_{pq,n_{1,i_1}} = (\varepsilon_{pn_{1,i_1}} + \varepsilon_{n_{1,i_1}q})/c$, and $\tau_{pq,n_{1,1},n_{1,2}} = (\varepsilon_{pn_{1,1}} + \varepsilon_{n_{1,1}n_{1,2}} + \varepsilon_{n_{1,2}q})/c$ are the waves' travel times through the link $T_p - R_q$, $T_p - s^{(n_{1,i_1})} - R_q$, and $T_p - s^{(n_{1,1})} - s^{(n_{1,2})} - R_q$, respectively, and c is the speed of light. $f_{T(R)m} = \upsilon_{T(R)}/\lambda$, $f_{T(R)m,1} = |\upsilon_{T(R)}\cos\gamma_{T(R)} - \upsilon_{ST}|/\lambda$, $f_{T(R)m,2} = |\upsilon_{T(R)}\cos\gamma_{T(R)} - \upsilon_{SR}|/\lambda$, and $f_{T(R)m,3} = \upsilon_{T(R)}/\lambda$ are the maximum Doppler frequency because the transmitter, receiver and some scatterers are moving. In addition, $\eta_{SB_{i_1}}$ and η_{DB} designate the contribution of the single- and double-bounced components to the total scattered power $\Omega_{pq}/(K+1)$, which satisfy $\sum_{i_1=1}^{I_1}\eta_{SB_{1,i_1}} + \eta_{DB} = 1$. For a low VTD, because most of the total power comes from the LoS component, the parameter K is large. What's more, since the number of moving scatterers is small, the static scatterers located on the first ellipse are allocated a large amount of power, and the proportion of the double-bounced components is smaller than that of the single reflected components. This can be express as $\eta_{SB_{1,3}} > \max\{\eta_{SB_{1,1}}, \eta_{SB_{1,2}}\} > \eta_{DB}$. In other side, the parameter K is small in a high VTD scene. What's more, since the amount of moving cars is large, the single-bounced components from the ellipse and two-ring models are allocated less energy than the double-bounced components from the two-ring models, which can be express as $\eta_{DB} > \max\{\eta_{SB_{1,1}}, \eta_{SB_{1,2}}, \eta_{SB_{1,3}}\}$. For other taps ($l' > 1$), the complex tap coefficient $h_{l',pq}(t)$ can be expressed as

$$h_{l',pq}(t) = h_{l',pq}^{SB}(t) + \sum_{i_2=1}^{I_2} h_{l',pq}^{DB_{i_2}}(t) \tag{2.3}$$

with

$$h_{l',pq}^{SB}(t) = \sqrt{\eta_{SB_{l',3}}\Omega_{pq}} \lim_{N_{l',3}\to\infty} \frac{1}{\sqrt{N_{l',3}}} \sum_{n_{l',3}=1}^{N_{l',3}} e^{j\phi^{(n_{l',3})}} e^{-2\pi f_c \tau_{pq,n_{l',3}}} e^{j2\pi t[f_{Tm}\cos(\alpha_T^{(n_{l',3})}-\gamma_T)+f_{Rm}\cos(\alpha_R^{(n_{l',3})}-\gamma_R)]} \tag{2.4a}$$

$$h_{l',pq}^{DB_1}(t) = \sqrt{\eta_{DB_{l',1}}\Omega_{pq}} \lim_{N_{1,1},N_{l',3}\to\infty} \frac{1}{\sqrt{N_{1,1}N_{l',3}}} \sum_{n_{1,1},n_{l',3}=1}^{N_{1,1},N_{l',3}} e^{j\phi^{(n_{1,1},n_{l',3})}} e^{-2\pi f_c \tau_{pq,n_{1,1},n_{l',3}}} e^{j2\pi t[f_{Tm,1}\cos(\alpha_T^{(n_{1,1})}-\gamma_T)+f_{Rm}\cos(\alpha_R^{(n_{l',3})}-\gamma_R)]} \tag{2.4a}$$

$$h_{l',pq}^{DB_2}(t) = \sqrt{\eta_{DB_{l',2}}\Omega_{pq}} \lim_{N_{l',3},N_{1,2}\to\infty} \frac{1}{\sqrt{N_{l',3}N_{1,2}}} \sum_{n_{l',3},n_{1,2}=1}^{N_{l',3},N_{1,2}} e^{j\phi^{(n_{l',3},n_{1,2})}} e^{-2\pi f_c \tau_{pq,n_{1,1},n_{1,2}}} e^{j2\pi t[f_{Tm}\cos(\alpha_T^{(n_{l',3})}-\gamma_T)+f_{Rm,2}\cos(\alpha_R^{(n_{1,2})}-\gamma_R)]} \tag{2.4b}$$

where $I_2 = 2$, $\tau_{pq,n_{l',3}} = (\varepsilon_{pn_{l',3}} + \varepsilon_{n_{l',3}q})/c$, $\tau_{pq,n_{l',3},n_{1,2}} = (\varepsilon_{pn_{l',3}} + \varepsilon_{n_{l',3}n_{1,2}} + \varepsilon_{n_{1,2}q})/c$, and $\tau_{pq,n_{1,1},n_{l',3}} = (\varepsilon_{pn_{1,1}} + \varepsilon_{n_{1,1}n_{l',3}} + \varepsilon_{n_{l',3}q})/c$ are the waves' travel times through the link $T_p - s^{(n_{l',3})} - R_q$, $T_p - s^{(n_{1,1})} - s^{(n_{l',3})} - R_q$, and $T_p - s^{(n_{l',3})} - s^{(n_{1,2})} - R_q$, respectively. The energy-related parameters η_{SB} and $\eta_{DB_{l',i_2}}$ satisfy $\sum_{i_2=1}^{I_2} \eta_{DB_{l',i_2}} + \eta_{SB} = 1$. For a low VTD, the static scatterers located on the ellipse are allocated a large amount of power, which indicates that $\eta_{SB_{l',3}} > \max\left\{\eta_{DB_{l',1}}, \eta_{DB_{l',2}}\right\}$. In addition, since the amount of moving vehicles is large under a high VTD, the single-bounced components from ellipse models are allocated less energy than the double-bounced components produced from the combination of one-ring and ellipse models, which can be express as $\min\left\{\eta_{DB_{l',1}}, \eta_{DB_{l',2}}\right\} > \eta_{SB_{l',3}}$. The AoDs and AoAs are discrete random variables, and they can be converted by the following widely used approximate relationships. For two rings model, $\alpha_R^{(n_{1,1})} = \pi - \frac{R_T}{D}\sin\alpha_T^{(n_{1,1})}$, $\alpha_T^{(n_{1,2})} = \pi - \frac{R_R}{D}\sin\alpha_R^{(n_{1,2})}$. For the multiple confocal ellipses model, $\sin\alpha_T^{(n_{l,3})} = \frac{b_l^2 \sin\alpha_R^{(n_{l,3})}}{a_l^2+f^2+2a_l f\cos\alpha_R^{(n_{l,3})}}$, $\cos\alpha_T^{(n_{1,3})} = \frac{2a_l f+(a_l^2+f^2)\cos\alpha_R^{(n_{l,3})}}{a_l^2+f^2+2a_l f\cos\alpha_R^{(n_{l,3})}}$. And in the case of the LoS components, $\alpha_T^{LoS} = 0$ and $\alpha_R^{LoS} = \pi$.

Since the scatterers distributed over the theoretical reference model are usually assumed to be infinity, one can use the continuous expressions of AoD $\alpha_T^{l,i}$ and AoA $\alpha_T^{l,i}$ to replace the discrete expressions $\alpha_T^{(n_{l,i})}$ and $\alpha_R^{(n_{l,i})}$, respectively. The azimuth angles $\alpha_T^{(n_{l,i})}$ and $\alpha_R^{(n_{l,i})}$ are characterized by using the von Mises PDF in this paper. The von Mises PDF [26] is defined as $f(\alpha) = exp\left[kcos(\alpha - \alpha_u)\right]/2\pi I_0(k)$, where $\alpha \in [-\pi, \pi)$, $I_0(\cdot)$ is the zeroth-order modified Bessel function of the first kind,

$\alpha_u \in [-\pi, \pi)$ accounts for the mean value of the angle α, and the real-valued parameter $k(k > 0)$ is designed to control the distribution of the angle α_u.

2.2.2 *Statistical Properties of V2V-MIMO Channel Model*

In this section, based on the research of literature on the statistical properties of narrowband V2V-MIMO reference models, some important channel statistical characteristics of the V2V-MIMO wideband channel model for non-isotropic scattering environment will be derived, including the STCF, DPSD, envelope LCR and AFD.

2.2.2.1 Space-Time Correction Function

The normalized space-time CF between two arbitrary complex tap coefficients $h_{pq}(t)$ and $h_{p'q'}(t)$ can be defined as $R_{pq,p'q'}(\delta_T, \delta_R, \tau) = \frac{\mathrm{E}[h^*_{pq}(t)h_{p'q'}(t+\tau)]}{\sqrt{\Omega_{pq}\Omega_{p'q'}}}$, where $E[\cdot]$ and $(\cdot)^*$ denote the statistical expectation operator and complex conjugate operation, respectively. It can be written as the superposition of the LoS, single- and double-bounced components. For the first tap,

$$R_{1,pq,p'q'}(\delta_T, \delta_R, \tau) = R^{LoS}_{1,pq,p'q'}(\delta_T, \delta_R, \tau) + R^{DB}_{1,pq,p'q'}(\delta_T, \delta_R, \tau) + \sum_{i_1=1}^{I_1} R^{SB_{1,i_1}}_{1,pq,p'q'}(\delta_T, \delta_R, \tau) \tag{2.5}$$

with

$$R^{LoS}_{1,pq,p'q'}(\delta_T, \delta_R, \tau) = \frac{K}{K+1} e^{-j\frac{2\pi}{\lambda}(\varepsilon_{pq}-\varepsilon_{p'q'})} e^{j2\pi\tau[f_{Tm}\cos(\alpha_T^{LoS}-\gamma_T)+f_{Rm}\cos(\alpha_R^{LoS}-\gamma_R)]} \tag{2.6a}$$

$$R^{SB_{1,i_1}}_{1,pq,p'q'}(\delta_T, \delta_R, \tau) = \frac{\eta_{SB_{1,i_1}}}{K+1} \int_{-\pi}^{\pi} f(\alpha^{1,i_1}_{T(R)}) e^{-j\frac{2\pi}{\lambda}[(\varepsilon_{pn_{1,i_1}}+\varepsilon_{n_{1,i_1}q})-(\varepsilon_{p'n_{1,i_1}}+\varepsilon_{n_{1,i_1}q'})]} e^{j2\pi\tau f_{Tm,i_1}\cos(\alpha^{1,i_1}_{T(R)}-\gamma_T)} e^{j2\pi\tau f_{Rm,i_1}\cos(\alpha^{1,i_1}_{R(T)}-\gamma_R)} d\alpha^{1,i_1}_{T(R)} \tag{2.6b}$$

$$R^{DB}_{1,pq,p'q'}(\delta_T,\delta_R,\tau) = \frac{\eta_{DB}}{K+1}\int_{-\pi}^{\pi}\int_{-\pi}^{\pi} f(\alpha_T^{1,1}) f(\alpha_R^{1,2}) \\ e^{-j\frac{2\pi}{\lambda}[(\varepsilon_{pn_{1,1}}+\varepsilon_{n_{1,2}q})-(\varepsilon_{p'n_{1,1}}+\varepsilon_{n_{1,2}q'})]} \\ e^{j2\pi\tau f_{Tm,1}\cos(\alpha_T^{1,1}-\gamma_T)} \\ e^{j2\pi\tau f_{Rm,2}\cos(\alpha_R^{1,2}-\gamma_R)} \mathrm{d}\alpha_T^{1,1}\mathrm{d}\alpha_R^{1,2}. \tag{2.6c}$$

For other taps,

$$R_{l',pq,p'q'}(\delta_T,\delta_R,\tau) = R^{SB}_{1,pq,p'q'}(\delta_T,\delta_R,\tau) + \sum_{i_2=1}^{I_2} R^{DB_{l',i_2}}_{l',pq,p'q'}(\delta_T,\delta_R,\tau) \tag{2.7}$$

with

$$R^{SB}_{l',pq,p'q'}(\delta_T,\delta_R,\tau) = \eta_{SB_{l',3}}\int_{-\pi}^{\pi} f(\alpha_R^{l',3}) \\ e^{-j\frac{2\pi}{\lambda}[(\varepsilon_{pn_{l',3}}+\varepsilon_{n_{l',3}q})-(\varepsilon_{p'n_{l',3}}+\varepsilon_{n_{l',3}q'})]} \\ e^{j2\pi\tau f_{Tm}\cos(\alpha_T^{l',3}-\gamma_T)} \\ e^{j2\pi\tau f_{Rm}\cos(\alpha_R^{l',3}-\gamma_R)} \mathrm{d}\alpha_R^{l',3} \tag{2.8a}$$

$$R^{DB_{l',1}}_{l',pq,p'q'}(\delta_T,\delta_R,\tau) = \eta_{DB_{l',1}}\int_{-\pi}^{\pi}\int_{-\pi}^{\pi} f(\alpha_T^{1,1}) f(\alpha_R^{l',3}) \\ e^{-j\frac{2\pi}{\lambda}[(\varepsilon_{pn_{1,1}}+\varepsilon_{n_{l',3}q})-(\varepsilon_{p'n_{1,1}}+\varepsilon_{n_{l',3}q'})]} \\ e^{j2\pi\tau f_{Tm,1}\cos(\alpha_T^{1,1}-\gamma_T)} \\ e^{j2\pi\tau f_{Rm}\cos(\alpha_R^{l',3}-\gamma_R)} \mathrm{d}\alpha_T^{1,1}\mathrm{d}\alpha_R^{l',3} \tag{2.8b}$$

$$R^{DB_{l',2}}_{l',pq,p'q'}(\delta_T,\delta_R,\tau) = \eta_{DB_{l',2}}\int_{-\pi}^{\pi}\int_{-\pi}^{\pi} f(\alpha_T^{l',3}) f(\alpha_R^{1,2}) \\ e^{-j\frac{2\pi}{\lambda}[(\varepsilon_{pn_{l',3}}+\varepsilon_{n_{1,2}q})-(\varepsilon_{p'n_{l',3}}+\varepsilon_{n_{1,2}q'})]} \\ e^{j2\pi\tau f_{Tm}\cos(\alpha_T^{l',3}-\gamma_T)} \\ e^{j2\pi\tau f_{Rm,2}\cos(\alpha_R^{1,2}-\gamma_R)} \mathrm{d}\alpha_T^{l',3}\mathrm{d}\alpha_R^{1,2}. \tag{2.8c}$$

2.2.2.2 Doppler Power Spectral Density

The Fourier Transform to the STCF can be used to describe the corresponding DPSD of the proposed V2V wideband model, which can be expressed as $S_{pl,qm}(f_D) = \int_{-\infty}^{+\infty} R_{pq,p'q'}(\tau)e^{-j2\pi f_D\tau}d\tau$, where f_D is the Doppler frequency.

2.2.2.3 Envelope LCR

Here, the LCR $L(r)$ is defined as the rate at which the signal envelope crosses a specified level r with a positive or negative direction. The LCR for V2V channels can be written as

$$
\begin{aligned}
L(r) =& \frac{2r\sqrt{K+1}}{\pi^{\frac{3}{2}}}\sqrt{\frac{b_2}{b_0} - \frac{b_1^2}{b_0^2}} \cdot e^{-K-(K+1)r^2} \\
& \int_0^{\pi/2} \cosh(2\sqrt{K(K+1)}r\cos\theta) \\
& [e^{-(\chi\sin\theta)^2} + \sqrt{\pi}\chi\sin\theta \cdot \mathrm{erf}(\chi\sin\theta)]d\theta
\end{aligned} \tag{2.9}
$$

where $\cosh(\cdot)$ and $\mathrm{erf}(\cdot)$ can be described as the hyperbolic cosine function and error function, and $\chi = \sqrt{\frac{Kb_1^2}{b_0b_2 - b_1^2}}$. First discuss the parameters b_0 of the first tap of the proposed model, which can be expressed as

$$
b_0 = b_0^{SB_{1,1}} + b_0^{SB_{1,2}} + b_0^{SB_{1,3}} + b_0^{DB} = \frac{1}{2(K+1)}. \tag{2.10}
$$

Similarly, the parameters b_1 and b_2 can be expressed as

$$
b_m = b_m^{SB_{1,1}} + b_m^{SB_{1,2}} + b_m^{SB_{1,3}} + b_m^{DB}, (m = 1, 2) \tag{2.11}
$$

with

$$
\begin{aligned}
b_m^{SB_{1,i_1}} =& \frac{\eta_{SB_{1,i_1}}}{2(K+1)}(2\pi)^m \int_{-\pi}^{\pi} f(\alpha_{T(R)}^{1,i_1})\{f_{Tm,i_1}[\cos(\alpha_T^{1,i_1} - \gamma_T) \\
& + f_{Rm,i_1}\cos(\alpha_R^{1,i_1} - \gamma_R)]\}^m d\alpha_{T(R)}^{1,i_1}
\end{aligned} \tag{2.12a}
$$

$$
\begin{aligned}
b_m^{DB} =& \frac{\eta_{DB}}{2(K+1)}(2\pi)^m \int_{-\pi}^{\pi}\int_{-\pi}^{\pi} f(\alpha_T^{1,1})f(\alpha_R^{1,2})\{f_{Tm,1} \\
& [\cos(\alpha_T^{1,1} - \gamma_T) + f_{Rm,2}\cos(\alpha_R^{1,2} - \gamma_R)]\}^m d\alpha_T^{1,1}d\alpha_R^{1,2}.
\end{aligned} \tag{2.12b}
$$

For other taps, the parameter b_0 can be obtained as

$$b_0 = b_0^{SB_{l',3}} + b_0^{DB_{l',1}} + b_0^{DB_{l',2}} = \frac{1}{2(K+1)}. \tag{2.13}$$

The parameters b_1 and b_2 can be expressed as

$$b_m = b_m^{SB_{l',3}} + b_m^{DB_{l',1}} + b_m^{DB_{l',2}}, \ (m = 1, 2) \tag{2.14}$$

with

$$\begin{aligned} b_m^{SB_{l',3}} =& \frac{\eta_{SB_{l',3}}}{2(K+1)}(2\pi)^m \int_{-\pi}^{\pi} f(\alpha_R^{l',3})\{f_{Tm}[\cos(\alpha_T^{l',3} - \gamma_T) \\ &+ f_{Rm}\cos(\alpha_R^{l',3} - \gamma_R)]\}^m \mathrm{d}\alpha_R^{l',3} \end{aligned} \tag{2.15a}$$

$$\begin{aligned} b_m^{DB_{l',1}} =& \frac{\eta_{DB_{l',1}}}{2(K+1)}(2\pi)^m \int_{-\pi}^{\pi}\int_{-\pi}^{\pi} f(\alpha_T^{1,1}) f(\alpha_R^{l',3})\{f_{Tm,1} \\ &[\cos(\alpha_T^{1,1} - \gamma_T) + f_{Rm,2}\cos(\alpha_R^{l',3} - \gamma_R)]\}^m \mathrm{d}\alpha_T^{1,1}\mathrm{d}\alpha_R^{l',3} \end{aligned} \tag{2.15b}$$

$$\begin{aligned} b_m^{DB_{l',2}} =& \frac{\eta_{DB_{l',2}}}{2(K+1)}(2\pi)^m \int_{-\pi}^{\pi}\int_{-\pi}^{\pi} f(\alpha_T^{l',3}) f(\alpha_R^{1,2})\{f_{Tm} \\ &[\cos(\alpha_T^{l',3} - \gamma_T) + f_{Rm,2}\cos(\alpha_R^{1,2} - \gamma_R)]\}^m \mathrm{d}\alpha_T^{l',3}\mathrm{d}\alpha_R^{1,2}. \end{aligned} \tag{2.15c}$$

The parameters b_0, b_1, and b_2 can be brought into the (2.9) to get the LCR of the proposed model.

2.2.2.4 Envelope AFD

The average time when the signal envelope $|h_{pq}(t)|$ stays below a certain level r is used to represent the signal envelope AFD $T(r)$. In the proposed model, the AFD can be written as

$$T(r) = \frac{1 - Q(\sqrt{2K}, \sqrt{2(K+1)r^2})}{L(r)} \tag{2.16}$$

where $Q(\cdot)$ is the Marcum Q function.

2.2.3 New 2D Wideband V2V-MIMO Channel Simulation Models

The reference model considers an infinite number of scatterers, but an infinite number of sinusoidal curves can not be realized in reality. Thus, designing a corresponding simulation model with limited complexity and capable of being implemented in practice is important. Therefore, a limited number of scatterers are consider in the simulation model, and this model is intended to use reasonable complexity to represent the expected channel statistical characteristics of the reference model as much as possible.

Based on the above discussion, the approximation of the channel statistical characteristics between the simulation model and the reference model depends on the sampling mode of the scatterers. When the scatterer's sample approaches the probability density function of the scatterer distribution in the reference model, the practicality of the simulation model will become stronger. In other words, it is important to find a way to design the sets of AoDs $\alpha_T^{(n_{l,i})}$ and AoAs $\alpha_R^{(n_{l,i})}$ of the simulation model to represent the desired channel statistical characteristics.

2.2.3.1 Deterministic Simulation Model

First, a deterministic simulation model is proposed, which requires constant parameters during simulation. In other words, based on the proposed model, the AoDs $\tilde{\alpha}_T^{(n_{l,i})}$ and AoAs $\tilde{\alpha}_R^{(n_{l,i})}$ have definite values for different simulation experiments. The AoDs and AoAs are designed as the follows.

1. First, define a new parameter $\tilde{\alpha}'^{(n_{l,i})}_{T(R)}$, which has the same environment parameters as $\tilde{\alpha}^{(n_{l,i})}_{T(R)}$ on a von Mises distribution.
2. Then, use the following method to get the set of $\left\{\tilde{\alpha}'^{(n_{l,i})}_{T(R)}\right\}_{n_{l,i}=1}^{N_{l,i}}$ as

$$\tilde{\alpha}'^{(n_{l,i})}_{T(R)} = F^{-1}_{T(R)}\left(\frac{n_{l,i} - 1/4}{N_{l,i}}\right), (n_{l,i} = 1, 2, \ldots, N_{l,i}) \tag{2.17}$$

   where $F^{-1}_{T(R)}$ denotes the inverse function of the von Mises CDF for $\tilde{\alpha}'^{(n_{l,i})}_{T(R)}$.
3. In order to ensure that the simulation models AoDs and AoAs in the range of $[-\pi, \pi)$, use the following mapping, that is

$$\tilde{\alpha}^{(n_{l,i})}_{T(R)} = \begin{cases} \tilde{\alpha}'^{(n_{l,i})}_{T(R)} + 2\pi & \tilde{\alpha}'^{(n_{l,i})}_{T(R)} < -\pi \\ \tilde{\alpha}'^{(n_{l,i})}_{T(R)} - 2\pi & \tilde{\alpha}'^{(n_{l,i})}_{T(R)} > \pi \\ \tilde{\alpha}'^{(n_{l,i})}_{T(R)} & else. \end{cases} \tag{2.18}$$

2.2.3.2 Stochastic Simulation Model

The deterministic simulation model is easy to implement and has a short simulation time. However, in an actual communication channel, the scatterers are not placed in a certain place like the proposed deterministic model. Therefore, this model can be transformed into a statistical simulation model if the phases or frequencies are random variables. Since the addition of random variables, the channel characteristics of the statistical simulation model change with each simulation trial, but will converge to the expectations of the model in a sufficient number of simulation trials. Similarly, the AoAs and AoDs designed by the statistical model can be expressed as follows.

1. First, propose a new random variable $\hat{\alpha}'^{(n_{l,i})}_{T(R)}$, which has the same environment parameters as $\hat{\alpha}^{(n_{l,i})}_{T(R)}$ on a von Mises distribution.
2. Then, use the following method to get the set of $\left\{\hat{\alpha}'^{(n_{l,i})}_{T(R)}\right\}_{n_{l,i}=1}^{N_{l,i}}$ as

$$\hat{\alpha}'^{(n_{l,i})}_{T(R)} = F^{-1}_{T(R)}\left(\frac{n_{l,i}+\theta-1/4}{N_{l,i}}\right), (n_{l,i}=1,2,\ldots,N_{l,i}). \tag{2.19}$$

   The parameters θ is independent random variable uniformly distributed on the interval $[-1/2, 1/2)$. Due to the introduction of random variable θ, the set of AoDs and AoAs varies with different simulation trial.
3. In order to ensure that the simulation model's AoDs and AoAs in the range of $[-\pi, \pi)$, use the following mapping, that is

$$\hat{\alpha}^{(n_{l,i})}_{T(R)} = \begin{cases} \hat{\alpha}'^{(n_{l,i})}_{T(R)} + 2\pi & \hat{\alpha}'^{(n_{l,i})}_{T(R)} < -\pi \\ \hat{\alpha}'^{(n_{l,i})}_{T(R)} - 2\pi & \hat{\alpha}'^{(n_{l,i})}_{T(R)} > \pi \\ \hat{\alpha}'^{(n_{l,i})}_{T(R)} & else. \end{cases} \tag{2.20}$$

2.2.3.3 Simulation Results and Analysis

In the following, the channel statistical characteristics between the proposed simulation models and reference model will be compared, and the usefulness of the two simulation models will be validated. The values of the parameters used for the numerical analysis are $D = 2f = 300\,\text{m}$, $R_T = R_R = 10\,\text{m}$, $\gamma_T = \gamma_R = 0$, $f_c = 5.9\,\text{GHz}$, $f_{Tm} = f_{Rm} = f_{T(R)m,3} = 570\,\text{Hz}$, $f_{T(R)m,1} = f_{T(R)m,2} = 360\,\text{Hz}$. Due to page constraints, only a three-tap model is considered, which the semi-major axis of the confocal multi-ellipse is assumed to be $a_l = \{160, 170, 180\}$.

Suppose the environment-parameters are $K = 3.8$, $k_T = 6.6$, $\mu_T = 12.8°$, $k_R = 8.3$, $\mu_R = 178.7°$, $k_E = 7.7$, $\mu_E = 31.3°$ for low VTD, and $K = 0.856$, $k_T = 0.6$, $\mu_T = 12.8°$, $k_R = 1.3$, $\mu_R = 178.7°$, $k_E = 8.5$, $\mu_E = 20.6°$ for high VTD. In the case of low VTD, for the first tap, it is assumed that the corresponding energy-parameters are $\eta_{SB_{1,1}} = 0.203$, $\eta_{SB_{1,2}}$=0.335, $\eta_{SB_{1,3}} = 0.411$, and $\eta_{DB} = 0.051$. For the l'th taps ($l' > 1$), assume that the corresponding energy parameters are $\eta_{SB_{l',3}} = 0.762$, $\eta_{DB_{l',1}} = 0.119$, and $\eta_{DB_{l',2}} = 0.119$. In the case of high VTD, for the first tap, the energy-parameters are set as $\eta_{SB_{1,1}} = 0.126$, $\eta_{SB_{1,2}} = 0.126$, $\eta_{SB_{1,3}} = 0.063$, and $\eta_{DB} = 0.685$. For the l'th taps ($l' > 1$), assume that the corresponding energy parameters are $\eta_{SB_{l',3}} = 0.088$, $\eta_{DB_{l',1}} = 0.456$, and $\eta_{DB_{l',2}} = 0.456$. The deterministic simulation model assumes that the effective scatterers' number are $N_{1,1} = N_{1,2} = N_{1,3} = N_{2,3} = N_{3,3} = 20$. For the statistical simulation model, assume the number of effective scatterers is $N_{1,1} = N_{1,2} = N_{1,3} = N_{2,3} = N_{3,3} = 20$ and the number of simulation experiments is $N_{stat} = 50$.

Figures 2.4 and 2.5 compare that when the transmitter and receiver are moving in the same and opposite directions, the DPSDs between the proposed reference and simulation models with different VTDs, respectively, where the (a) shows the

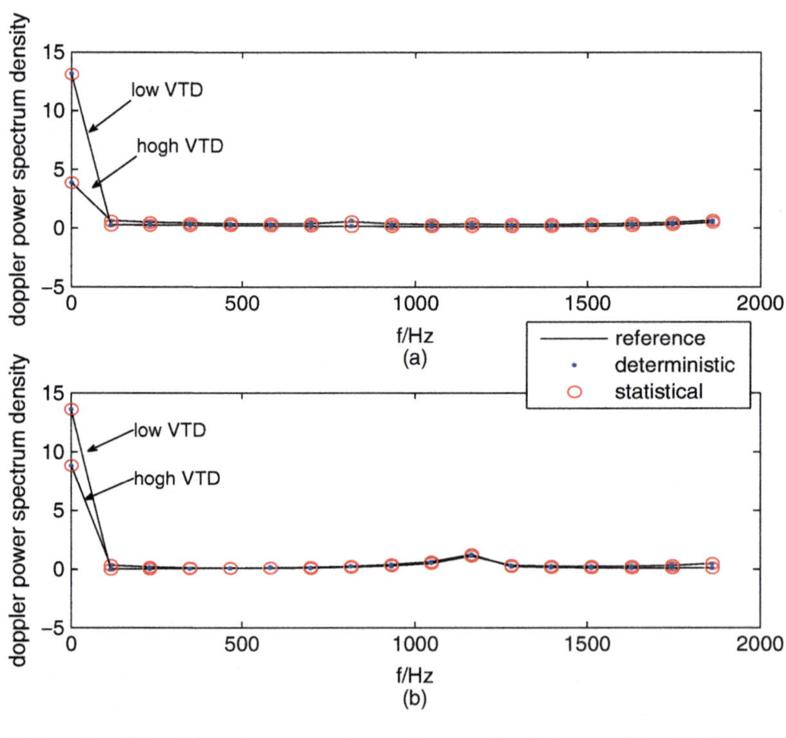

Fig. 2.4 Doppler PSDs of the reference model and the two simulation models with different VTDs for the same direction of movement of the Tx and Rx under two different tap scenarios: (**a**) First tap and (**b**) Second tap

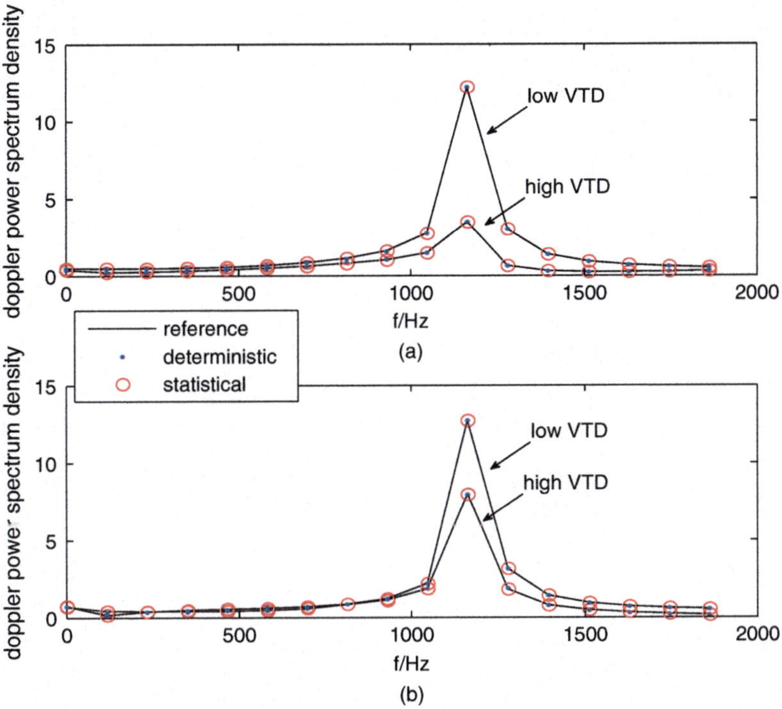

Fig. 2.5 Doppler PSDs of the reference model and the two simulation models with different VTDs for the opposite direction of movement of the Tx and Rx under two different tap scenarios: (**a**) First tap and (**b**) Second tap

first tap and (b) shows the second tap. It can be seen from the image that there is a high approximation between the reference model and the simulation model, which further proves the usefulness of the proposed wideband model. At the same time, the VTDs also have some effect on the DPSD. It can be seen that the more uniform the DPSD distribution, the higher the corresponding VTD. This is because that the received power is mainly allocated to the LoS components and the static scatterers in the case of low VTDs, whereas the power is mainly allocated to the mobile scatterer in the case of high VTDs. In addition, the spatial separation leads to the fluctuation of DPSD, resulting in more uniform distribution of the PSD. Similarly, when the speed direction of Tx and Rx are same or opposite, Figs. 2.6 and 2.7 show the LCR between the reference and simulation model with different VTDs, where the (a) and (b) show the first and second tap, respectively. In addition, they show that the corresponding LCR increases as the VTD increases. This phenomenon can be explained that mobile vehicles are allocated most of the energy because of the high VTD.

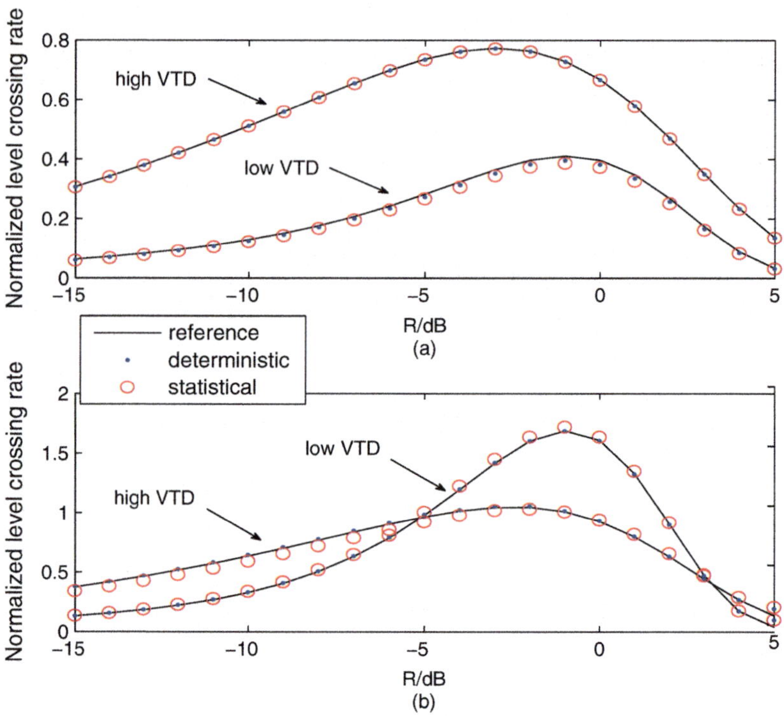

Fig. 2.6 Envelope LCR of the reference model and the two simulation models with different VTDs for the same direction of movement of the Tx and Rx under two different tap scenarios: (**a**) First tap and (**b**) Second tap

Figure 2.8 shows the STCF of the proposed models with different VTDs when the speed direction of Tx and Rx are same and opposite. It can be seen that the VTD affects the STCF of the model. The space-time correlation decreases as the VTD increases. This phenomenon can be interpreted as the spatial diversity of the channel increases as the VTD increases, and the correlation decreases. At the same time, these phenomena show that the STCF of the two simulation models proposed in this paper can achieve a high approximation to the reference model within a range of time delay, which further proves the correctness and practicability of our proposed model.

In addition, although the deterministic simulation model is easy to be implemented and can be achieved with a short simulation time, the results have shown that the matching degree between reference model and statistical simulation model is higher than that of deterministic model in case of a small amount of scatterers. Therefore, the statistical simulation model is more suitable for the estimation of channel characteristics under the premise that the computational complexity is similar.

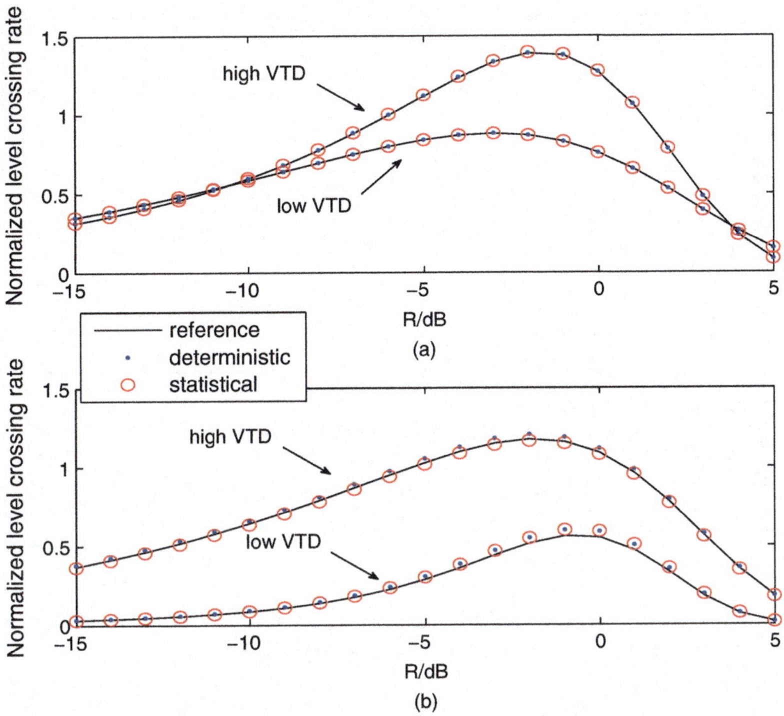

Fig. 2.7 Envelope LCR of the reference model and the two simulation models with different VTDs for the opposite direction of movement of the Tx and Rx under two different tap scenarios: (**a**) First tap and (**b**) Second tap

2.3 New Features Due to 5G

Since both the Tx and Rx are in motion with low elevation antennas, as well as many highly mobile clusters/scatterers surrounding the Tx and Rx, vehicular communication channels express significantly different and unique properties. Several survey papers [3, 27] have discussed and summarized vehicular communication channel measurements, which are all below 6 GHz with 1–4 antennas. One key feature of 5G is transmission rate: as envisioned by the communications industry, 5G should support more than 10 times the data rate of 4G, i.e., 1 Gb/s. There are mainly two approaches to improve the transmission rate in wireless systems: (1) Improve the spectrum efficiency; and (2) Increase the transmission bandwidth. For 5G, there are many potential technologies for spectrum efficiency improvement. Among those, the most widely adopted and channel-pertinent one is massive MIMO, which typically has more than 100 antennas. To increase the spectrum bandwidth, 5G first considered millimeter wave (mmWave) band with GHz bandwidth. Therefore, new technologies such as mmWave and massive MIMO

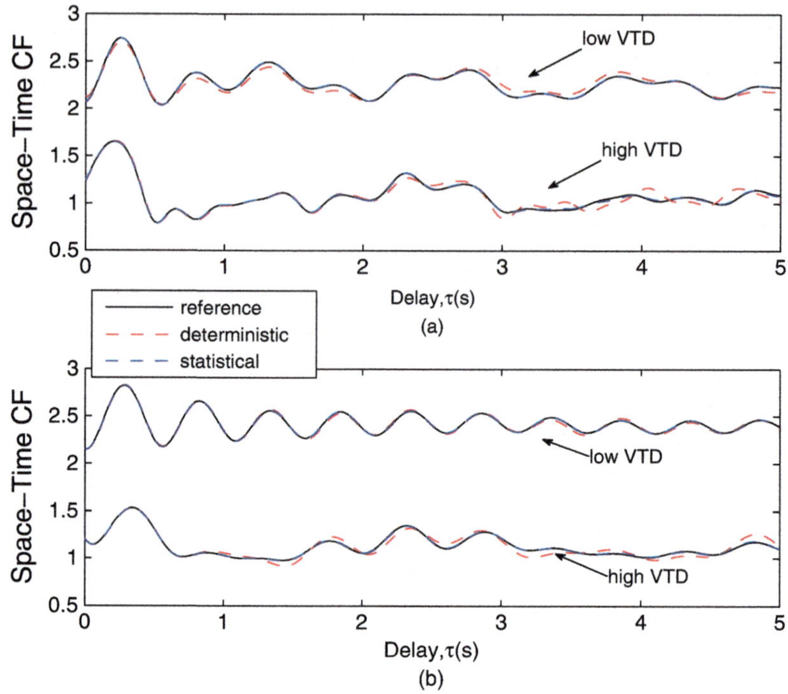

Fig. 2.8 Space-Time correlation function of the reference model and the two simulation models with low and high VTDs when the Tx and Rx move (**a**) in the same direction and (**b**) in the opposite direction

lead to new unique characteristics of 5G vehicular communications (5G-VehC) channels, thus demanding channel measurements that are specifically designed for such technologies. Next, 5G-VehC channel measurements from mmWave and massive MIMO perspectives will be reviewed briefly.

2.3.1 mmWave Perspective

Compared to vehicular communication channel measurements below 6 GHz, vehicular mmWave channel measurements are still at infancy [28]. In the presence, mainstream mmWave measurements mostly deploy directional horn antennas. Rappaport et al. [29] carried out vehicular mmWave channel measurement campaigns in representative environments, which revealed that mmWave channels exhibit moderate multipath spread and are thus suitable for wireless transmission, while the blocking of buildings may significantly affect the transmission quality. Hur et al. [30] uses vehicular polarized antenna to measure the path loss and polarization properties at 28 and 73 GHz frequency bands. Both of the aforementioned measurements considered the V2I scenario. In [31], V2V communication channel measurements were conducted in static parking lot scenarios, where the 60 GHz band with non line of sight vehicle blockage. Recently, Samsung for the first time conducted dynamic V2V channel measurements, in which the 28 GHz band was considered under vehicle speed of 100 km/h [32].

The aforementioned measurements provided the basic large- and small-scale characteristics of vehicular mmWave channels such as higher path loss, larger shadowing, higher delay resolution, and fewer multi paths in one cluster, and carried out primary discussions on the feasibility of applying mmWave in VCN. However, relative to the mmWave measurements in conventional cellular scenarios, vehicular mmWave channel measurements are still at very primary stage and urgently call for additional measurement campaigns under various scenarios and at different frequency bands.

2.3.2 Massive MIMO Perspective

In comparison with conventional MIMO, massive MIMO channel measurements are way more difficult and complicated. At present, the mainstream measurement approach for massive MIMO is to use virtual antenna arrays mainly below 6 GHz, with the emphasis of revealing the characteristics and variations of the channel under different antenna patterns. Several massive MIMO channel measurement campaigns tailored for conventional cellular scenarios under 6 GHz [33] revealed two unique properties of massive MIMO channels: spherical wavefront and spatial non-stationarity. Massive MIMO has a large number of antennas, which results in that the distance between the Tx (cluster) and Rx may not be larger than the Rayleigh distance. In this case, the effect of spherical wavefront is significant and the conventional plane wavefront assumption cannot be used anymore. Because of the large antenna array involved in massive MIMO, certain clusters are no longer observable over the entire array. This means that each antenna element on the large array may have its own set of clusters. This leads to the non-stationarity in the spatial domain, i.e., channel statistics vary along the large antenna array.

As aforementioned, mmWave renders massive MIMO more feasible with a more reasonable form factor. A few measurement campaigns have been carried out for mmWave massive MIMO channels and have demonstrated that the two aforementioned unique properties are more evident, i.e., more obvious spherical wavefront and more severe spatial non-stationarity compared with massive MIMO in the 6 GHz band and below. It is worth noting that the aforementioned massive MIMO measurement campaigns are all designed for conventional cellular scenarios. So far there has been no measurement campaign available for 5G frequency band (2–85 GHz) massive MIMO vehicular communication channels. Although the channel characteristics revealed by these existing measurement campaigns could throw some clues to massive MIMO vehicular communication channels, it is still urgently necessary to set up massive MIMO measurement campaign for VCN scenarios in order to unveil the unique channel properties therein.

2.4 3D Space-Time-Frequency Non-stationarity

The complex high-speed mobile environment in VCN gives rise to clear time varying properties in vehicular channels. Not only that the channels vary over time, but also the statistical properties of the channel also vary over time, thus leading to the temporal non-stationarity, i.e., the non-stationarity in the time domain, which is the importantly unique property of vehicular channels [27]. As mentioned before, the large number of antennas in massive MIMO render its corresponding channel characteristics changes along the antenna array dimension, thus leading to spatial non-stationarity, i.e., the non-stationarity in the space domain, which is the importantly unique property of massive MIMO channels [33]. In addition, mmWave channels occupy higher frequency band and ultra wideband, and hence exhibit significant frequency dispersion. In other words, the different frequency components of the same signal experiences different propagation parameters, or even different propagation mechanisms, leading to frequency-dependent channel characteristics. These give rise to the frequency non-stationarity, i.e., the non-stationarity in the frequency domain, that is the importantly unique property of mmWave channels [28]. For 5G-VehC channels, the non-stationarity is simultaneously present in all three domains. More importantly, the rapid time variation, massive MIMO, and ultra-high frequency and bandwidth associated with 5G-VehC channels, together with their interactions, will aggravate these non-stationarities. Therefore, 5G-VehC channels exhibit more severe and unique 3D space-time-frequency non-stationarity. This subsection will first elaborate the underlying physical mechanism of the 3D non-stationarity and then introduce possible non-stationarity modeling approaches.

Mathematically speaking, the non-stationarity of the channel in a certain domain implies that the channel statistical characteristics are varying in this domain. As a result, 3D space-time-frequency non-stationarity implies that the channel statistics varies with space, time, and frequency. Note that a wireless channel has three dual domains as shown in Fig. 2.9, i.e., time t v.s. Doppler f_D, frequency f_c v.s. delay τ, and space x v.s. angle/direction Ω [36]. This means that the time domain and Doppler domain features reflect the same phenomenon of the channel, but from difference perspectives/domains [36]. Similarly, the frequency/space domain and delay/angle domain also reflect the same channel properties from different perspectives. As a result, channel non-stationarity in the time/frequency/space domain is equivalent to channel correlated scattering in the Doppler/delay/angle domain. The widely used WSSUS assumption actually refers to the wide sense stationarity in both time and frequency domains.

To better illustrate the physical mechanisms underlying 3D non-stationarity, the time, delay, and space domain perspectives will be took. This is because, in these three domains, clusters/scatterers could be concretely described as shown

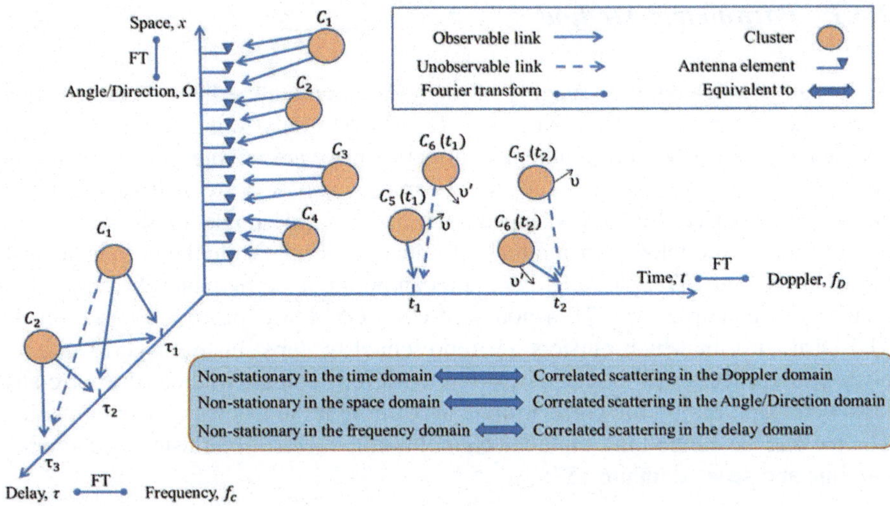

Fig. 2.9 Physical mechanisms underlying 3D non-stationarity in time, delay, and space. Note that unobservable links are omitted in the space domain for clarity

in Fig. 2.9. It is clear from Fig. 2.9 that due to the fast channel time-variation, the clusters/scatterers appearance and disappearance can happen along the time axis t, which results in the non-stationarity in the time domain. Similarly, due to the large scale antenna array, the clusters/scatterers appearance and disappearance can occur along the antenna array axis x. This leads to the non-stationarity in the space domain. Due to the high delay resolution, the clusters/scatterers in different delays could be correlated as shown in Fig. 2.9. This results in the correlated clustering/scattering in the delay domain, i.e., the non-stationarity in the frequency domain. More importantly, one can easily observe from Fig. 2.9 that the fast channel time-variation will enhance the clusters/scatterers appearance and disappearance along the antenna array axis x, and the large antenna array will enhance the clusters/scatterers appearance and disappearance along the time axis t. This means that time non-stationarity and space non-stationarity will mutually enhance each other. The same conclusion could be made between time-frequency non-stationarity and space-frequency non-stationarity. As a result, 3D space-time-frequency non-stationarity exhibit stronger non-stationarity compared to the non-stationarity in individual space domain, time domain, or frequency domain. Based on these detailed and illustrative description of non-stationarity, how to reasonably model non-stationarity will be discussed. To date, there have been some works modeling channel non-stationarity using seemingly differing approaches. In lieu of the above analysis and interpretation, all these approaches could be categorized as follows: parametric method, geometric method, and hybrid method.

2.4.1 Parametric Method

As its name indicated, this category of methods completely relies on mathematical approaches to model non-stationarity. In NGSM, only this method could be adopted. Specifically, pure mathematical approaches are adopted in modeling the clusters/scatterers appearance and disappearance, and/or correlated scattering. In general, the clusters/scatterers appearance and disappearance in the time/space domain can be modeled by using the birth-death process, while the correlated scattering in the delay/Doppler/angle domain can be modeled by using the correlation matrix. For example, in [27], a non-WSSUS NGSM was proposed based on the TDL structure, in which clusters are represented by taps. In this NGSM, the tap birth-death process was modeled by discrete Markov chains, whereas the correlated taps were captured by the correlation matrix through the Cholesky decomposition. The parametric method is simple but cannot mimic the drift of clusters/scatterers in the time and space domains [37].

2.4.2 Geometric Method

The geometric method models the non-stationarity naturally by directly describing the dynamic scattering geometry environments, including the movement of the Tx, Rx, and clusters/scatterers. Any modeling approaches via the description of the underlying scattering geometry, such as GBDM and GBSM, can use the geometric method to model non-stationarity. Specifically, this category of methods describes the types, locations, and moving traces of clusters/scatters, Tx, and Rx, at various levels of details, to naturally model non-stationarity. For example, GBDM introduced in [3] describes the dynamic scattering geometry in great detail, whereas GBSM introduced in [37] adopts simplified statistical approach to describe dynamic scattering geometry, and therefore naturally model the non-stationarity. Although geometric method could better describe the drift of clusters/scatterers in the time and space domains, but this comes at very high complexity [37].

2.4.3 Hybrid Method

Considering the trade-off between complexity and accuracy, hybrid method can properly combine the aforementioned parametric and geometric methods. This category of methods flexibly adopts mathematical tools to reasonably reduce the complexity in modeling non-stationarity by describing the dynamic scattering geometry environments. In general, the hybrid method first models the clusters/scatterers appearance and disappearance in the time/space domain by using a mathematical birth-death process, and then over time updates clusters/scatterers parameters, e.g., locations, delays, powers, angles, etc., with the help of the description of the dynamic scattering geometry environments. For example, a non-stationary RS-GBSM [34] was developed based on the TDL structure. This RS-GBSM first mimics the taps birth-death process by using discrete Markov chains, and then derives the time-varying AoA and AoD based on the regular scattering geometry. A non-stationary IS-GBSM [35] first models the cluster birth-death process by adopting discrete Markov chains, and then updates clusters location, delay, power, AoA, and AoD based on the irregular scattering geometry.

2.5 Challenges and Open Issues

The challenges discussed in this section can be considered as guidelines for setting up future measurement campaigns, as well as developing more realistic 5G vehicular communications channel models.

2.5.1 Channel Measurement

The high-speed mobile environments in VCN make 5G-VehC channel measurements very challenging in several counts. (1) Establishment of measurement platforms: The widely adopted virtual antenna array (VAA) based massive MIMO measurement approach is no longer feasible. This is because the VAA measurement should ensure that the single antenna for measurement is shifted to all needed positions within the channel coherence time. However, the dynamic fast time-varying environments associated with 5G-VehC massive MIMO result in very short channel coherence time, making it impossible to complete the VAA based massive MIMO measurements. As a result, 5G-VehC massive MIMO measurements may have to resort to real antenna array (RAA) based approaches. This will significantly increase the measurement cost as well as necessitate accurate and complicated large antenna array calibration. If mmWave band is also considered, then one would have to adopt approaches involving horn antenna RAA that are accurately controlled electronically, thus further increasing the difficulty and cost of channel measurement. (2) Measurement of complex characteristics: As mentioned before, 3D non-stationarity is a unique property of 5G-VehC channels. Though measurements have revealed existence of such 3D non-stationarity from various perspectives, they are not sufficient to support accurate modeling of the 3D non-stationarity, such as the entangled birth-death processes of clusters/scatterers in time and space domains. In summary, 5G-VehC channel measurements should be able to accurately describe the delay, angle spread, and time variation of the channel components. All these urgently call for the establishment of joint space-time-frequency 5G-VehC channel measurement strategies.

2.5.2 Channel Modeling

The high-speed mobile environments in VCN make 5G-VehC channel modeling very challenging in several counts. (1) Extremely limited measurement data As aforementioned, the uniqueness of the 5G-VehC scenario makes the establishment of the measurement platform very difficult. In such circumstances, one can attempt to construct 5G-VehC communication scenarios of interest using calibrated GBDM, as opposed to actual measurements, to obtain relatively reasonable channel properties, and then combine with GBSM to establish proper channel models. Such a modeling approach combining GBDM and GBSM initiates a new angle to effectively solve the above challenge. (2) 3D non-stationarity modeling: Existing modeling works regarding non-stationarity either model the non-stationarity in a single domain, e.g., in the time/space domain, or model the non-stationarity in a 2D domain, e.g., in the time-space/time-frequency domain. While the modeling of 3D non-stationarity is quite limited, how to properly model 3D non-stationarity is still an open problem. (3) Partitioned spatial correlation modeling: As mentioned before, 5G-VehC massive MIMO exhibits obvious 3D non-stationarity, hence leading to regionalized property along the antenna array in terms of the channel spatial correlation. However, current massive MIMO is still using the traditional linear spatial correlation modeling in conventional MIMO channels, lacking research in partitioned spatial correlation modeling and leaving this a blank area to be filled.

References

1. M. Boban, T. T. V. Vinhoza, M. Ferreira, J. Barros, and O. K. Tonguz, "Impact of vehicles as obstacles in vehicular ad hoc networks," *IEEE J. Sel. Areas Commun.*, vol. 29, no. 1, pp. 15–28, Jan. 2011.
2. T. Abbas, F. Tufvesson, K. Sjoberg, and J. Karedal, "Measurement based shadow fading model for vehicle-to-vehicle network simulations," *IEEE Trans. Veh. Technol.*, arXiv: 1203.3370v3, June 2013.
3. C.-X. Wang, X. Cheng, and D. I. Laurenson, "Vehicle-to-vehicle channel modeling and measurements: recent advances and future challenges," *IEEE Commun. Mag.*, vol. 47, no. 11, pp. 96–103, Nov. 2009.
4. G. Acosta, K. Tokuda, and M. A. Ingram, "Measured Joint Doppler-Delay Power Profiles for Vehicle-to-Vehicle Communications at 2.4 GHz," in *Proc. IEEE GLOBE-COM 04*, Dallas, TX, Nov. 2004, pp. 3813–17.
5. I. Sen and D. W. Matolak, "Vehicle-vehicle Channel Models for the 5-GHz Band," *IEEE Trans. on Intell. Transp. Syst.*, vol. 9, no. 2, pp. 235–245, Jun. 2008.
6. A. Paier et al., "Characterization of Vehicle-to-Vehicle Radio Channels from Measurements at 5.2 GHz," *IEEE Wireless Personal Commun.*, June 2008.
7. L. Cheng et al., "Mobile Vehicle-to-Vehicle Narrowband Channel Measurement and Characterization of the 5.9 GHz Dedicated Short Range Communication (DSRC) Frequency Band," *IEEE JSAC*, vol. 25, no. 8, Oct. 2007, pp. 1501–16.
8. G. Acosta and M. A. Ingram, "Six time- and frequency-selective empirical channel models for vehicular wireless LANs," *IEEE Veh. Technol. Mag.*, vol. 2, no. 4, pp. 4–11, Dec. 2007.

9. O. Renaudin, V.-M. Kolmonen, P. Vainikainen, and C. Oestges, "Car-to-car channel models based on wideband MIMO measurements at 5.3 GHz," in *Proc. EuCAP*, Berlin, Germany, Mar. 2009, pp. 635–639.
10. J. Maurer, T. Fgen, and W. Wiesbeck, "A ray-optical channel model for vehicle-to-vehicle communication," in *Proc. in Physics: Fields, Networks, Computational Methods, and Systems in Modern Electrodynamics, P. Russer and M. Mongiardo*, Eds. Berlin, Germany: Springer, 2004, pp. 243–254.
11. J. Karedal, F. Tufvesson, N. Czink, A. Paier, C. Dumard, T. Zemen, C. F. Mecklenbrauker and A. F. Molisch, "A geometry-based stochastic MIMO model for vehicle-to-vehicle communications," *IEEE Trans. Wireless Commun.*, vol. 8, no. 7, pp. 3646–3657, Jul. 2009.
12. A. S. Akki and F. Haber, "A statistical model for mobile-to-mobile land communication channel," *IEEE Trans. Veh. Technol.*, vol. VT-35, no. 1, pp. 2–7, Feb. 1986.
13. M. Patzold, B. O. Hogstad and N. Youssef, "Modeling, analysis, and simulation of MIMO mobile-to-mobile fading channels," *IEEE Trans. Wireless Commun.*, vol. 7, no. 2, pp. 510–520, Feb. 2008.
14. A. G. Zajic and G. L. Stuber, "Space-Time Correlated Mobile-to-Mobile Channels: Modelling and Simulation," *IEEE Trans. Veh. Technol.*, vol. 57, no. 2, pp. 715–726, Mar. 2008.
15. X. Cheng, C. X. Wang, D. I. Laurenson, S. Salous and A. V. Vasilakos, "An adaptive geometry-based stochastic model for non-isotropic MIMO mobile-to-mobile channels," *IEEE Trans. Wireless Commun.*, vol. 8, no. 8, pp. 4824–4835, Sept. 2009.
16. X. Cheng, C. X. Wang, B. Ai and H. Aggoune, "Envelope Level Crossing Rate and Average Fade Duration of Nonisotropic Vehicle-to-Vehicle Ricean Fading Channels," *IEEE Trans. on Intell. Transp. Syst.*, vol. 15, no. 1, pp. 62–72, Feb. 2014.
17. A. G. Zajic and G. L. Stuber, "Three-Dimensional Modeling, Simulation, and Capacity Analysis of Space-Time Correlated Mobile-to-Mobile Channels," *IEEE Trans. Veh. Technol.*, vol. 57, no. 4, pp. 2042–2054, Jul. 2008.
18. A. S. Akki, "Statistical properties of mobile-to-mobile land communication channels," *IEEE Trans. Veh. Technol.*, vol. 43, no. 4, pp. 826–831, Nov. 1994.
19. A. G. Zaji, G. L. Stber, T. G. Pratt, and S. Nguyen, "Envelope level crossing rate and average fade duration in mobile-to-mobile fading channels," in *Proc. IEEE ICC*, Beijing, China, May 2008, pp. 4446–4450.
20. A. G. Zaji, G. L. Stber, T. G. Pratt, and S. Nguyen, "Wideband MIMO mobile-to-mobile channels: Geometry-based statistical modeling with experimental verification," *IEEE Trans. Veh. Technol.*, vol. 58, no. 2, pp. 517–534, Feb. 2009.
21. X. Cheng, C.-X.Wang, D. I. Laurenson, and A. V. Vasilakos, "Second order statistics of non-isotropic mobile-to-mobile Ricean fading channels," in *Proc. IEEE ICC*, Dresden, Germany, Jun. 2009, pp. 1–5.
22. A. G. Zajic and G. L. Stuber, "Three-dimensional modeling and simulation of wideband MIMO mobile-to-mobile channels," *IEEE Trans. Wireless Commun.*, vol. 8, no. 3, pp. 1260–1275, Mar. 2009.
23. A. G. Zajic and G. L. Stuber,"3-D Simulation Models for Wideband MIMO Mobile-to-Mobile Channels," in *Proc. IEEE MCC*, 2007, pp. 1–5.
24. X. Cheng, Q. Yao, M. Wen, C. X. Wang, L. Y. Song and B. L. Jiao, "Wideband Channel Modeling and Intercarrier Interference Cancellation for Vehicle-to-Vehicle Communication Systems," *IEEE J. on Sel. Area.Comm.*, vol. 31, no. 9, 2013, pp. 434–448.
25. Y. Li, X. Cheng and N. Zhang, "Deterministic and Stochastic Simulators for Non-Isotropic V2V-MIMO Wideband," *IEEE China Comm.*, 2018.
26. A. Abdi and M. Kaveh, "A space-time correlation model for multielement antenna systems in mobile fading channels," *IEEE J. on Sel. Area.Comm.*, vol. 20, no. 3, pp. 550–560, Apr. 2002.
27. D. W. Matloak, "Channel modeling for vehicle-to-vehicle communications," *IEEE Communications Magazine*, vol. 46, no. 5, pp. 76–83, May 2008.

28. J. Zhang, P. Tang, L. Tian, Z. Hu, T. Wang, and H. Wang, "6–100 GHz Research Progress and Challenges for Fifth Generation (5G) and Future Wireless Communication from Channel Perspective," *IEEE SCIENCE CHINA Information Sciences*, vol. 60, no. 8, pp. 1–16, Aug. 2017.
29. T. S. Rappaport, G. R. MacCartney, M. K. Samimi, and S. Sun, "Wideband millimetre-wave propagation measurements and channel models for future wireless communication system design," *IEEE Transactions on Wireless Communications*, vol. 63, no. 9, pp. 3029–3056, Sept. 2015.
30. S. Hur, S. Baek, B. Kim, Y. Chang, A. F. Molisch, T. S. Rappaport, K. Haneda, and J. Park, "Proposal on millimetre-wave channel modelling for 5G cellular system," *IEEE Journal of Selected Topic in Signal Processing*, vol. 10, no. 3, pp. 454–469, Mar. 2016.
31. E. B. Dor, T. S. Rappaport, Y. Qiao, and S. J. Lauffenburger, "Millimeter-wave 60 GHz outdoor and vehicle AOA propagation measurements using a broadband channel sounder," in *Proc. IEEE Globe Communications Conference*, GLOBECOM 2011, Houston, USA, Dec. 2011.
32. J. Gozalvez, "Samsung electronics sets 5G speed record at 7.5 Gb/s," *IEEE Vehicular Magazine*, vol. 10, no. 1, pp. 12–16, Jan. 2015.
33. J. Zhang, Z. Zheng, Y. Zhang, J. Xi, X. Zhao, and G. Gui,"3D MIMO for 5G NR: Several Observations from 32 to Massive 256 Antennas Based on Channel Measurement," *IEEE Communication Magazine*, Mar. 2018.
34. Y. Yuan, C.-X. Wang, Y. He, M. M. Alwakeel, and H. Aggoune, "3D wideband non-stationary geometry-based stochastic models for non-isotropic MIMO vehicle-to-vehicle channels," *IEEE Transactions on Wireless Communications*, vol. 14, no. 12, pp. 6883–6895, Dec. 2015.
35. A. Ghazal, Y. Yuan, C.-X. Wang, Y. Zhang, Q. Yao, Y. Yuan, H. Zhou, and W. Duan, "A non-stationary IMT-A MIMO channel model for high-mobility wireless communication systems," *IEEE Transactions on Wireless Communications*, vol. 16, no. 4, pp. 2057–2068, Apr. 2017.
36. X. Yin and X. Cheng, Propagation Channel Characterization, Parameter Estimation, and Modeling for Wireless Communications, John Wiley Sons Inc, 2016
37. A. F. Molisch, F. Turfvesson, J. Karedal and C. F. Mecklenbrauker, "A Survey On Vehicle-To-Vehicle Propagation Channels," *IEEE Wireless Commun.*, vol. 16, no. 6, pp. 12–22, 2009.

Chapter 3
Wireless-Vehicle Combination: Advanced PHY Techniques in VCN

Along with the fast development of both the automobile industry and mobile communication systems, the combination of vehicles and wireless communications has been a burgeoning trend and is of essential significance in safety and mobility applications in vehicular networks. However, as described in Chap. 2, compared with traditional (quasi-)static communications, wireless channels in vehicular communications are more complex with large Doppler and evident non-stationarity, which makes 5G-enabled vehicle-to-everything (V2X) communications face rigorous challenges. Hence, in this chapter, in order to combat the severe channel conditions and achieve high data rate, we first explore some advanced PHY techniques for VCN system design, which can effectively exploit the benefits of OFDM and MIMO in high mobility vehicular networks. Then, shifting from combating channel to exploiting channel, we also discuss the potential PHY directions for the next leap.

3.1 PHY Techniques in VCN

Currently, dedicated short range communications (DSRC) is a widely applied solution for vehicular communications, which is developed based on the IEEE 802.11p standard [1]. In the PHY layer, IEEE 802.11p follows the same frame structure, modulation schemes, and training sequences as IEEE 802.11a but with doubled time-domain parameters to fit the outdoor vehicular scenarios, where OFDM is employed as the core PHY technique. Although DSRC can provide two-way short-(or medium-)range reliable vehicle-to-infrastructure (V2I) and vehicle-to-vehicle (V2V) communications, the achievable data rate is very limited that may not satisfy many emerging vehicular applications (e.g., video/image transmission and autonomous driving). Hence, LTE-vehicle (LTE-V) [2] or cellular vehicle-to-everything (C-V2X) [3], which combines the advantages of both cellular

X. Cheng et al., *5G-Enabled Vehicular Communications and Networking*, Wireless Networks, https://doi.org/10.1007/978-3-030-02176-4_3

communications and ad-hoc connectivity among vehicles, has been regarded as the rising star in the current development of vehicular communications. Compared with DSRC, in the PHY layer, LTE-V and C-V2X employ more advanced techniques, such as MIMO and cooperative schemes, which can significantly improve the reliability and efficiency of vehicular communications, leading to high data rate and low latency.

No matter in current DSRC or in future C-V2X, OFDM and MIMO are expected to be the core PHY techniques for vehicular communications. However, due to the high mobility in vehicular communications, the application of OFDM is facing severe ICI that may destroy the orthogonality of OFDM subcarriers and result in unreliable transmissions. Thus, how to effectively alleviate the effect of ICI has been regarded as the most important problem for practical OFDM-based vehicular communications. In addition, due to the fast time-varying channels in vehicular communications, channel estimation becomes very challenging, which may affect the accuracy of the obtained CSI and thus degrade the MIMO efficiency. Then, how to fully exploit the benefits of MIMO techniques in vehicular communications consists of a big challenge.

Hence, in the following, considering the challenges for OFDM and MIMO in vehicular networks that include severe ICI and rapidly time-varying fading channels, we introduce some essential and recent PHY techniques which can effectively address these facing challenges in future 5G-enabled VCN, including novel ICI cancellation methods, index modulated (IM-)OFDM, differential spatial modulation, and energy harvesting relaying. The content structure of this chapter is given in Fig. 3.1.

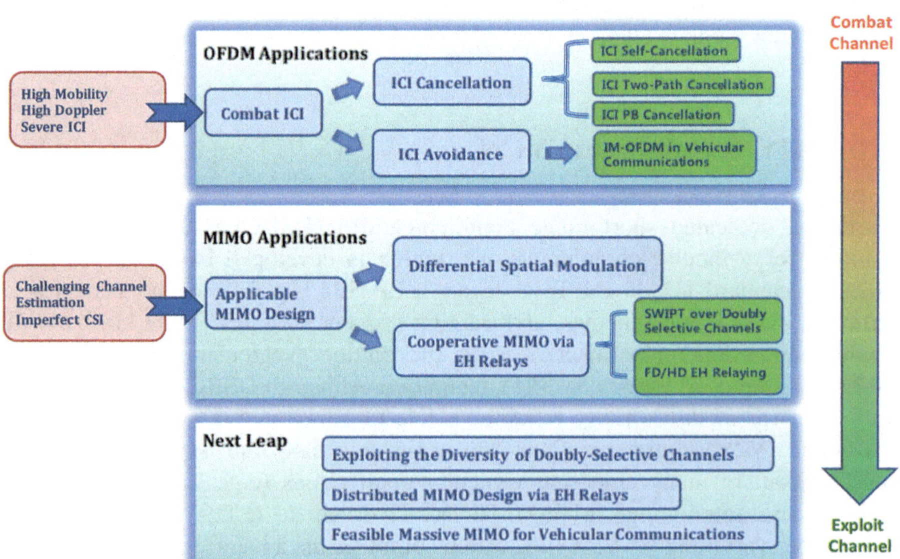

Fig. 3.1 Content structure of 5G-VCN PHY techniques

3.2 ICI Cancellation for OFDM

The performance of OFDM systems is determined by the specific orthogonal properties among OFDM subcarriers. The subcarriers' orthogonality can be easily destroyed by carrier frequency offset (CFO), leading to ICI. ICI problem becomes more serious for V2X communications due to the high Doppler frequency caused by the fast movement of the transmitter (Tx), receiver (Rx), and scatterers. This section will focus on the fundamental issue on the proper use of OFDM technique in V2X networks: how to simply and effectively mitigate ICI? In this section, we will discuss some effective ICI cancellation schemes suitable for vehicular communications and further provide a novel constant phase rotation-aided (CPRA) method that can be combined with most ICI cancellation schemes to improve the ICI mitigation performance.

3.2.1 ICI Cancellation Schemes

In the literature, many existing works have contributed to effective and feasible ICI cancellation in OFDM systems, which can be summarized as:

- ICI Self-Cancellation [4]: Insert mirror subcarriers into a transmitter OFDM symbol and exploit the symmetric features of the mirrored subcarriers to alleviate the ICI. According to the mirror mapping rules, the ICI self-cancellation schemes can categorized into two classes, that is, mirror symbol repetition (MSR) and mirror conjugate symbol repetition (MCSR).
- Two-Path Transmission ICI Cancellation [5, 7, 8]: Utilize data repetition across two concatenated OFDM blocks for ICI cancellation.
- Precoding-Based ICI Cancellation (PBC) [6]: Combine the advantages of ICI self-cancellation and two-path ICI cancellation, which can be easily implemented into practical communication systems.

3.2.1.1 ICI Self-Cancellation

ICI self-cancellation method can efficiently combat ICI by employing data repetition within one OFDM symbol interval. As depicted in Fig. 3.2a, the additional processes for the ICI self-cancellation method with respect to the normal OFDM procedure are the ICI self-cancellation module, including the ICI canceling modulation before the inverse fast Fourier transform (IFFT) operation at the Tx and the ICI canceling demodulation after the fast Fourier transform (FFT) operation at the Rx [9].

For the ICI canceling modulation, the input modulated symbols $\{X\}$ are first grouped into several transmit blocks each consisting of $\frac{N}{2}$ modulated symbols, which are then mapped onto the N subcarriers using one-to-two mapping rule:

$$\widetilde{X}_{\mathcal{M}\{k\}} = \begin{cases} X_k,\ k = 0, \ldots, \frac{N}{2} - 1 \\ \mathcal{O}\left(X_{k-\frac{N}{2}}\right),\ k = \frac{N}{2}, \ldots, N - 1 \end{cases} \tag{3.1}$$

where $\{\widetilde{X}_k\}_{k=0}^{N-1}$ are the real transmitted symbols on the OFDM subcarriers, $\mathcal{M}$ is a mapping set whose elements are chosen according to the specific mapping criterion, such as adjacent mapping, symmetric mapping, and mirror mapping, and $\mathcal{O}(x)$ is defined as the mapping operation which reflects the relationship between the two symbols with the same information to be mapped on the subcarrier pair. In ICI self-cancellation methods, conversion and conjugate relationships are commonly utilized and can be represented in (3.1) as $\mathcal{O}(x) = -x$ and $\mathcal{O}(x) = x^*$, respectively. Therefore, there are six ICI self-cancellation schemes with the combination of the three mapping criteria and two mapping operations: adjacent symbol repetition (ASR) [10], adjacent conjugate symbol repetition (ACSR) [11], symmetric symbol repetition (SSR) [12], symmetric conjugate symbol repetition (SCSR) [13], mirror symbol repetition (MSR), and mirror conjugate symbol repetition (MCSR) [4]. Taking the mirror mapping criterion with the conjugate mapping operation, i.e., MCSR scheme, as an example, we have $\mathcal{M} = \{1, \ldots, \frac{N}{2} - 1, \emptyset, N - 1, \ldots, \frac{N}{2} + 1, \emptyset\}$ with $\emptyset$ denoting invalid mapping. Therefore, the final transmitted symbols in the frequency domain after the ICI self-cancellation module is $\widetilde{X}_1 = \widetilde{X}_{N-1}^* = X_0, \widetilde{X}_2 = \widetilde{X}_{N-2}^* = X_1, \ldots, \widetilde{X}_{N/2-1} = \widetilde{X}_{N/2+1}^* = X_{N/2-2}, \widetilde{X}_0 = \widetilde{X}_{N/2} = 0$ [4].

After the ICI cancelling modulation, the received signals on subcarrier m and its corresponding mapped subcarrier pair m' ($m' = m + 1$ for the adjacent mapping criterion, $m' = N - m - 1$ for the symmetric mapping criterion, or $m' = N - m$ for the mirror mapping criterion) will carry on the same data symbol. This signal redundancy makes it possible to further improve the ICI mitigation performance through a combination technique, which can be realized as

$$\begin{aligned} \widehat{X}_m = & \frac{\mathcal{O}\left(H^*_{\mathcal{M}\left\{m+\frac{N}{2}\right\},\mathcal{M}\left\{m+\frac{N}{2}\right\}} Y_{\mathcal{M}\left\{m+\frac{N}{2}\right\}}\right)}{\left|H_{\mathcal{M}\{m\},\mathcal{M}\{m\}}\right|^2 + \left|H_{\mathcal{M}\left\{m+\frac{N}{2}\right\},\mathcal{M}\left\{m+\frac{N}{2}\right\}}\right|^2} \\ & + \frac{H^*_{\mathcal{M}\{m\},\mathcal{M}\{m\}} Y_{\mathcal{M}\{m\}}}{\left|H_{\mathcal{M}\{m\},\mathcal{M}\{m\}}\right|^2 + \left|H_{\mathcal{M}\left\{m+\frac{N}{2}\right\},\mathcal{M}\left\{m+\frac{N}{2}\right\}}\right|^2} \end{aligned} \tag{3.2}$$

where $H_{m,k} = \frac{1}{N} \sum_{l=1}^{L} F_l(m-k) e^{-j\frac{2\pi lk}{N}}$, $m = 0, \ldots, \frac{N}{2} - 1$, $F_l(z) = \sum_{n=0}^{N-1} h_l^{(n)} e^{-j\frac{2\pi nz}{N}}$, and $h_l^{(n)}$ is the l-th sample of the time-varying channel impulse response at time instant $\frac{nT_s}{N}$ with T_s denoting the symbol duration.

3.2.1.2 Two-Path Transmission ICI Cancellation

Unlike the ICI self-cancellation method where the data is repeated within one OFDM block, the two-path transmission method transmits the data copies in two concatenated OFDM blocks, which are usually referred to as two independent paths separated by time division multiplexing (TDM), as shown in Fig. 3.2b. Similar to the ICI self-cancellation method, by applying the combination technique on the received signals from both paths at the Rx, the ICI generated from one path can be significantly mitigated by that generated from the other path.

Currently, parallel cancellation (PC) [7] and conjugate cancellation (CC) [8] schemes have been known as typical two-path cancellation schemes. Taking the CC scheme as an example, the first path follows the normal OFDM procedure, while the second path adopts a conjugate of the normal OFDM signal both before the transmission and after the reception, as illustrated in Fig. 3.2b, without considering the $e^{j\phi}$ and $-e^{j\phi}$ modules. Finally, the received signals at the m-th subcarrier for both paths are combined as (3.3) by setting $\phi = 0$, where the effect of noise is ignored. More recently, to further improve the ICI mitigation performance of CC scheme, the PRCC scheme has been proposed by using PRA method that adds an artificial phase rotation of ϕ for both paths at the Tx in the CC scheme, as shown in Fig. 3.2b. However, the performance of PRCC scheme is highly dependent on the

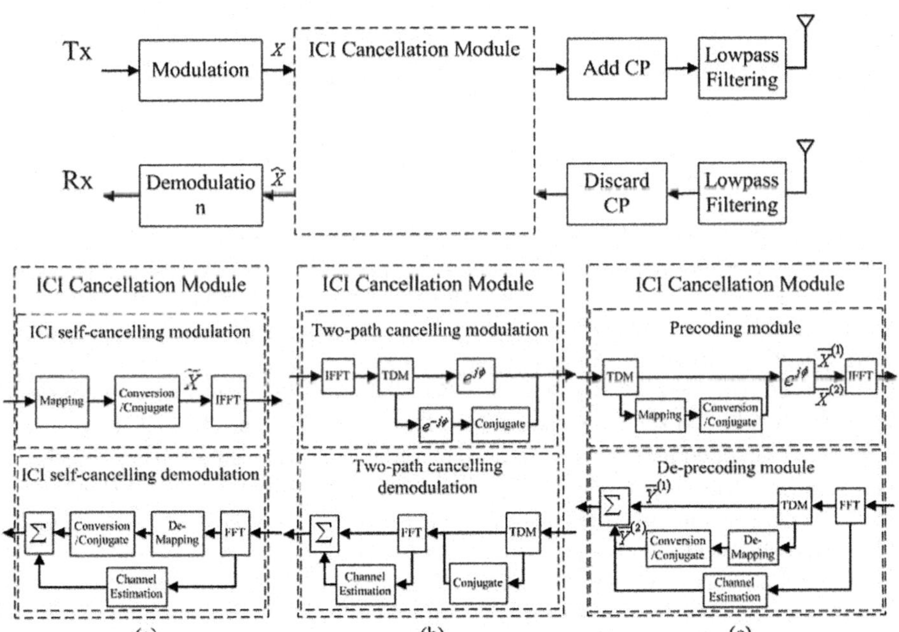

Fig. 3.2 Block diagram of an OFDM transceiver with (**a**) ICI self-cancellation module, (**b**) PRCC two-path cancellation module, and (**c**) PBC with PRA method module

accuracy of the chosen phase rotation, which is related to the real frequency offset experienced at the Rx and thus needs a precise estimation at the Rx and an error-free feedback link to the Tx. For example, the phase rotation adopted at the Tx is derived as $\phi = -\pi\hat{\varepsilon}\left(1 - \frac{1}{N}\right)$, where $\hat{\varepsilon}$ is the feedback estimated frequency offset from the Rx. The PRA method used in the PRCC scheme aims to make the phases of the ICI coefficients for the both paths have an opposite polarity, so that they can cancel out each other when the combination of the both paths is implemented. Therefore, the combination of received signals at the m-th subcarrier for both paths can be expressed as

$$\widehat{X}_m = \frac{e^{j\phi}\left|H_{m,m}\right|^2 + e^{-j\phi}\left|H_{-m,-m}\right|^2}{\left|H_{m,m}\right|^2 + \left|H_{-m,-m}\right|^2} X_m + \frac{\sum\limits_{k=0,k\neq m}^{N-1}\left(e^{j\phi}H_{m,m}^{*}H_{m,k} + e^{-j\phi}H_{-m,-m}H_{-m,-k}^{*}\right)X_k}{\left|H_{m,m}\right|^2 + \left|H_{-m,-m}\right|^2}. \tag{3.3}$$

It has been shown in [14] that due to the application of the phase rotation-aided method (PRA) method, the PRCC scheme outperforms most existing ICI cancellation schemes under an ideal assumption that the accurate phase rotation is obtained at the Tx. Note that the PRA method proposed in the PRCC scheme can be also applicable to any ICI self-cancellation and two-path cancellation schemes for further improving their ICI mitigation performance.

3.2.1.3 Precoding-Based ICI Cancellation (PBC)

From the above descriptions, it can be found that the essence of the ICI self-cancellation and two-path transmission schemes is the proper design of the redundant signals associated with the combination technique. Specifically, the redundant signals for ICI self-cancellation schemes are introduced within one OFDM block, while those for two-path transmission schemes are in two concatenated OFDM blocks. Compared with the normal OFDM procedure, the complexity increase of ICI self-cancellation schemes is marginal. However, when the ICI self-cancellation schemes are applied into the IEEE 802.11p based vehicular communication systems, performance degradation will appear as the configuration of the OFDM block, which includes not only data symbols but also pilot symbols, seriously breaks the mapping criterion to the data symbols. Fortunately, this problem can be successfully avoided by the application of the two-path transmission schemes due to no additional operation within one OFDM block. The two separate OFDM transceivers, however, will significantly increase the realization complexity of two-path transmission schemes and as a consequence limit the practical use of the two-path transmission schemes. This motivates a new type of ICI cancellation method proposed in [6], which combines both the advantages of the ICI self-cancellation and two-path cancellation schemes and meanwhile is suitable for practical vehicular communication systems.

In [6], the authors discovered a relationship between ICI self-cancellation schemes and two-path transmission schemes and proved that the CC scheme is actually another implementation manner of the MCSR scheme. Motivated by this, a new type of ICI cancellation method was then proposed where the main additional operations compared with normal OFDM systems have been integrated inside the precoding and de-precoding modules, as shown in Fig. 3.2c, without considering the $e^{j\phi}$ module and thus named as PBC scheme. The PBC scheme involves the redundant signals by properly applying the mapping criterion and mapping operation in two concatenated OFDM blocks separated by TDM. For the precoding module, one OFDM block input $\{X_k\}_{k=0}^{N-1}$ will become two OFDM blocks output $\{\overline{X}_k^{(1)}\}_{k=0}^{N-1}$ and $\{\overline{X}_k^{(2)}\}_{k=0}^{N-1}$, where the first OFDM block $\{\overline{X}_k^{(1)}\}_{k=0}^{N-1}$ is equal to the input OFDM block, i.e., $\overline{X}_k^{(1)} = X_k$ and the second OFDM block $\{\overline{X}_k^{(2)}\}_{k=0}^{N-1}$ is obtained by following the mapping rule $\overline{X}_{\mathcal{M}\{k\}}^{(2)} = \mathcal{O}(X_k)$. Similar to (3.1), $\mathcal{M}$ is a mapping set whose elements are chosen according to the specific mapping criterion, such as adjacent mapping with $\mathcal{M} = \{1, 2, \ldots, N-1, 0\}$, symmetric mapping with $\mathcal{M} = \{N-1, N-2, \ldots, 0\}$, and mirror mapping with $\mathcal{M} = \{0, N-1, N-2, \ldots, 1\}$, and $\mathcal{O}(x)$ is defined as the mapping operation which is either conversion operation or conjugate operation. Therefore, similar to ICI self-cancellation schemes, in total there are six PBC schemes, namely precoding based ASR (PB-ASR), PB-ACSR, PB-SSR, PB-SCSR, PB-MSR, and PB-MCSR. Finally, the combination of the received signals at the m-th subcarrier for both OFDM blocks can be derived as

$$\widehat{X}_m = \frac{H_{m,m}^* \overline{Y}_m^{(1)} + \mathcal{O}\left\{H_{\mathcal{M}\{m\},\mathcal{M}\{m\}}^* \overline{Y}_{\mathcal{M}\{m\}}^{(2)}\right\}}{\left|H_{m,m}\right|^2 + \left|H_{\mathcal{M}\{m\},\mathcal{M}\{m\}}\right|^2}. \tag{3.4}$$

where $\{\overline{Y}_m^{(1)}\}_{m=0}^{N-1}$ and $\{\overline{Y}_m^{(2)}\}_{m=0}^{N-1}$, respectively, are the received signals corresponding to the transmitted two OFDM blocks in the frequency domain.

By properly applying the mapping criterion and mapping operation into two concatenated OFDM blocks rather than into one OFDM block as in ICI self-cancellation schemes, the proposed PBC schemes can be directly implemented into practical vehicular communication systems without exhibiting any performance degradation. Moreover, unlike two-path cancellation schemes that design the two concatenated OFDM blocks after the IFFT at the Tx and before the FFT at the Rx, as illustrated in Fig. 3.2b, the design of the two OFDM blocks in the PBC scheme is integrated in the precoding module before the IFFT at the Tx and the de-precoding module after the FFT at the Rx, as shown in Fig. 3.2c. This indicates that the precoding and de-precoding operations can be accomplished via a little software programming effort without modifying the hardware of present OFDM systems as previous two-path cancellation schemes have to do. Therefore, by properly taking the both advantages of ICI self-cancellation schemes and two-path cancellation schemes, the proposed PBC schemes can be easily implemented into practical vehicular communication systems and have no any performance degradation.

3.2.2 Constant Phase Rotation-Aided (CPRA) Method

The PRA method can be effectively used in the ICI self-cancellation, two-path transmission ICI cancellation, and PBC schemes to further improve the ICI mitigation performance. The key issue on the implementation of the PRA method is how to obtain the accurate artificial phase rotation at the Tx. In general, the implementation of the PRA method needs a powerful frequency offset estimator at the Rx and an error-free feedback link from the Rx to the Tx, which will significantly increase the system complexity and reduce the system real-time performance. Due to the complicated moving environments of practical vehicular communication systems, the frequency offset caused by Doppler should be a spectrum largely spread in a wide range rather than a single value. Therefore, a precise estimation of the frequency offset in vehicular systems will not only be a serious burden at the Rx but also hardly be obtained. Moreover, the PRA method is derived based on a single-valued frequency offset and thus may not be suitable for practical vehicular communication systems.

Taking this into consideration, in [6], a CPRA method was proposed where the artificial phase rotation is constant and irrelevant to the frequency offset. Therefore, the implementation of the proposed CPRA method is simple and only needs a $e^{j\phi}$ module with a given constant optimal phase rotation ϕ at the Tx. More importantly, the constant phase rotation is designed over a given frequency offsets range rather than based on a single-valued frequency offset. The carrier-to-interference-ratio (CIR) is used to design the CPRA method. Therefore, the basic design rule is that the CIR of any ICI cancellation scheme with the designed CPRA method should overwhelm the CIR of the normal OFDM systems and meanwhile approach the CIR of the ICI cancellation scheme with the PRA method for any given fixed frequency offset. By taking the PRPB-MCSR as an example, this section details the design procedure of the CPRA method. Considering a single-valued frequency offset $\Delta f = f_d$ where f_d denotes any Doppler shift in the Doppler spectrum of $h_l^{(n)}$, the CIRs of normal OFDM systems and constant phase rotation PB-MCSR (CPRPB-MCSR) with a phase rotation ϕ can be expressed as

$$C_{OFDM} = \frac{\mathrm{sinc}^2(\varepsilon)}{1 - \mathrm{sinc}^2(\varepsilon)} \tag{3.5}$$

and

$$C_{CPRPB-MCSR} = \frac{\mathrm{sinc}^2(\varepsilon)\cos^2\left(\pi\varepsilon\left(1 - \frac{1}{N}\right) + \phi\right)}{\frac{1}{2} + \frac{1}{2}\mathrm{sinc}(2\varepsilon)\, X \text{ - } \mathrm{sinc}^2(\varepsilon)\, Y} \tag{3.6}$$

respectively, where $X = \cos\left(2\pi\varepsilon\left(1-\frac{1}{N}\right)+2\phi\right)$, $Y = \cos^2\left(\pi\varepsilon\left(1-\frac{1}{N}\right)+\phi\right)$, $\mathrm{sinc}\,(x) = \sin(\pi x)/\pi x$, and the normalized frequency offset $\varepsilon = \Delta f T_s$. The CIR of the PRPB-MCSR can be obtained directly from (3.6) by setting $\phi = -\pi\varepsilon\left(1-\frac{1}{N}\right)$ as

$$C_{PRPB-MCSR} = \frac{\mathrm{sinc}^2(\varepsilon)}{\frac{1}{2}+\frac{1}{2}\mathrm{sinc}\,(2\varepsilon)-\mathrm{sinc}^2(\varepsilon)} \tag{3.7}$$

which is actually the maximum of (3.6). Note that for the PRPB-MCSR, the artificial phase rotation is relevant to the normalized frequency offset ε.

3.2.2.1 The Range of Feasible Solutions of ϕ

Dividing the nominator and denominator in (3.6) by $\cos^2\left(\pi\varepsilon\left(1-\frac{1}{N}\right)+\phi\right)$, it is observed that the only difference among (3.5)–(3.7) lies in the denominator. Therefore, to guarantee that the CIR in (3.6) is larger than that in (3.5) for a given frequency offset range $\varepsilon \in [\varepsilon_{\min}, \varepsilon_{\max}]$, where $\varepsilon_{\min} > 0$ and $\varepsilon_{\max} \leq 0.5$ are the minimum and maximum normalized frequency offsets, respectively, then

$$\psi(\varepsilon,\phi) < 1,\ \ \psi(\varepsilon,\phi) = \frac{1+\mathrm{sinc}\,(2\varepsilon)\cos\left(2\pi\varepsilon\left(1-\frac{1}{N}\right)+2\phi\right)}{2\cos^2\left(\pi\varepsilon\left(1-\frac{1}{N}\right)+\phi\right)}$$
$$\approx \frac{1+\mathrm{sinc}\,(2\varepsilon)\cos(2\pi\varepsilon+2\phi)}{1+\cos(2\pi\varepsilon+2\phi)} \tag{3.8}$$

where $\varepsilon \in [\varepsilon_{\min}, \varepsilon_{\max}]$ and the approximation is derived based on the reasonable assumption of a large value of N. Calculating the above inequality, one can obtain that $\cos(2\pi\varepsilon+2\phi)(1-\mathrm{sinc}\,(2\varepsilon)) > 0$. Resorting to the property $\mathrm{sinc}\,(2\varepsilon) < 1$, feasible solutions of ϕ can be obtained over the range as $[-\frac{\pi}{4}-\pi\varepsilon_{\min}+k\pi, \frac{\pi}{4}-\pi\varepsilon_{\max}+k\pi]$ with k being an integer. Without loss of generality, in the sequel, ϕ is constrained in its principal value interval $\phi \in \left[-\frac{\pi}{4}-\pi\varepsilon_{\min}, \frac{\pi}{4}-\pi\varepsilon_{\max}\right]$.

3.2.2.2 Derivation of Optimal ϕ

Based on the basic design rule, the optimal ϕ should be chosen such that the CIR for any ε drawn from the possible frequency offset range experienced at the Rx approaches the saturation level, i.e., the CIR of the PRPB-MCSR. Therefore, if we define a metric $\mathscr{D}$ which characterizes the difference between the CIR of the

CPRPB-MCSR and the CIR of the PRPB-MCSR for a given ε, the optimal ϕ will be figured out by minimizing the accumulated differences of CIRs with ε spreading from $\varepsilon_{\min}$ to $\varepsilon_{\max}$

$$\phi_{opt} = \min\{\rho(\phi)\}, \ \rho(\phi) = \int_{\varepsilon_{\min}}^{\varepsilon_{\max}} \mathscr{D}\left(C_{CPRPB-MCSR}(\varepsilon, \phi), C_{PRPB-MCSR}(\varepsilon)\right) d\varepsilon. \tag{3.9}$$

where $\mathscr{D}\left(C_{CPRPB-MCSR}(\varepsilon, \phi), C_{PRPB-MCSR}(\varepsilon)\right) = |C_{PRPB-MCSR}(\varepsilon) - C_{CPRPB-MCSR}(\varepsilon, \phi)|^2$ is the square error between these two CIRs and $\phi \in \left[-\frac{\pi}{4} - \pi\varepsilon_{\min}, \frac{\pi}{4} - \pi\varepsilon_{\max}\right]$. The integration range $[\varepsilon_{\min}, \varepsilon_{\max}]$ should be determined by the transceiver performance and communication environments as the former determines the level of phase noise and frequency mismatch between the received signal and local oscillator, and the latter determines the level of Doppler spectrum. The optimal ϕ_{opt} can be readily obtained by minimizing $\rho(\phi)$ through computer numerical calculating. Especially, if $[\varepsilon_{\min}, \varepsilon_{\max}] = [0, 0.5]$, the feasible solutions of ϕ are reduced to one element, $\phi = -\frac{\pi}{4}$, which actually turns out to be the optimal phase rotation.

Therefore, it can be clearly obtained that the phase rotation of the PRA method is the optimal one for any given single-valued ε, while the phase rotation of the CPRA method is the suboptimal one for any given single-valued ε but is the optimal one over a given frequency offset range $[\varepsilon_{\min}, \varepsilon_{\max}]$. Since the Doppler spectrum for vehicular communication systems in practical results in a wider spread frequency offset range rather than a single-valued frequency offset, compared with the PRA method, the CPRA method should express better performance and more importantly has very small implementation complexity.

In the following, we evaluate the introduced ICI cancellation schemes in practical V2V communication channels, including opposite direction (OD) low and high vehicular traffic density scenarios (i.e., OD-LVTD and OD-HVTD). In the simulations, the number of OFDM subcarriers is $N = 64$, the CP length is $Lp = 16$, both short and long training preambles consisting of 2 OFDM symbols are transmitted before data frame for the signal synchronization, frequency offset estimation, and channel estimation, 4 pilot symbols are multiplexed with the transmitted data in the frequency domain for phase tracking. Least squares (LS) estimator is used for the channel estimation.

Figure 3.3 shows the BER performance of the practical V2V communication system without considering pilot symbols under OD-LVTD and OD-HVTD channels when either ICI self-cancellation schemes or PBC schemes are applied. The symbols in all above cases are drawn from QPSK constellation. As expected, the BER performance of ICI self-cancellation schemes is similar to that of the PBC schemes. Moreover, it can be seen that the SSR and the MCSR outperform the ASR. This is because that the larger frequency separation in the subcarrier mapping criterion for the SSR and MCSR helps collect the additional frequency diversity and thus enhances their capability in the ICI cancellation.

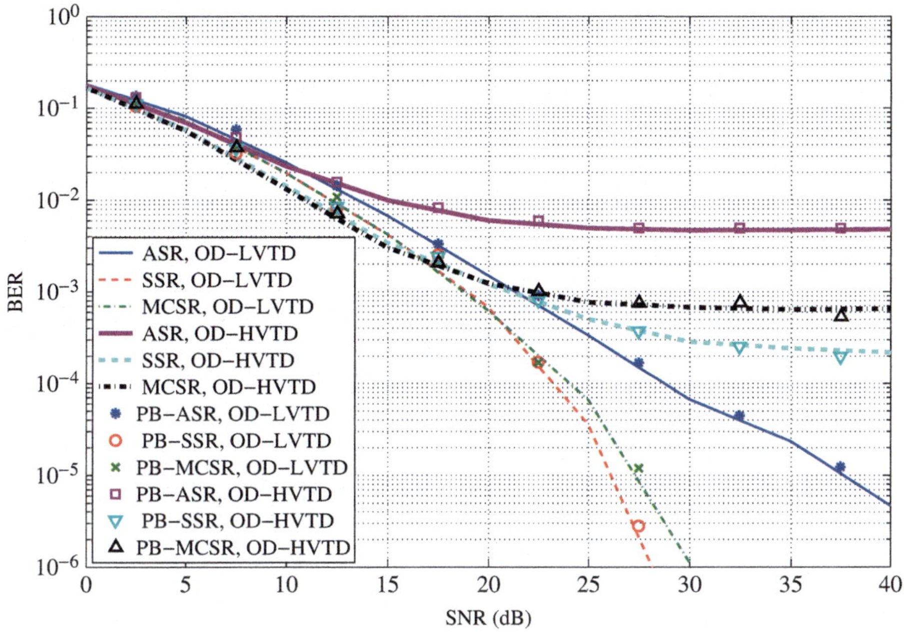

Fig. 3.3 BER performance of different ICI self-cancellation schemes and PBC schemes without considering pilots under OD-LVTD and OD-HVTD V2V channels

Figure 3.4 shows the performance of the implementation of both the ICI self-cancellation schemes and the PBC schemes into practical V2V communication systems under OD-LVTD and OD-HVTD channels. The V2V communication system without using any ICI cancellation schemes is also considered as a baseline. To guarantee the same spectral efficiency, we use BPSK modulation for the case without ICI cancellation schemes and the QPSK modulation for the case using ICI cancellation schemes. To show the impact of pilot symbols, equally-spaced pilot symbols with the number $Np = 4$ and $Np = 8$ are considered. From Fig. 3.4, it can be clearly found that the presence of pilot symbols significantly deteriorates the performance of ICI self-cancellation schemes since the pilot symbols will break the mapping criterion. Moreover, the performance degradation increases with the increase of the number of pilot symbols.

Figure 3.5 illustrates the BER performance of the V2V communication system with the PB-MCSR, PRPB-MCSR, and CPRPB-MCSR schemes under OD-LVTD and OD-HVTD channels. For the PRPB-MCSR, the phase rotation is obtained by using the estimated normalized frequency offset $\hat{\varepsilon}$ from the preamble. While for the CPRPB-MCSR, the constant phase is obtained based on the normalized frequency offset range $[\hat{\varepsilon} - \varpi, \hat{\varepsilon} + \varpi]$ with $\varpi = 0.02$ in this simulation. As shown in Fig. 3.5, the CPRPB-MCSR achieves the best performance, validating the better performance of the CPRA method over the PRA one. Note that the performance gap between the

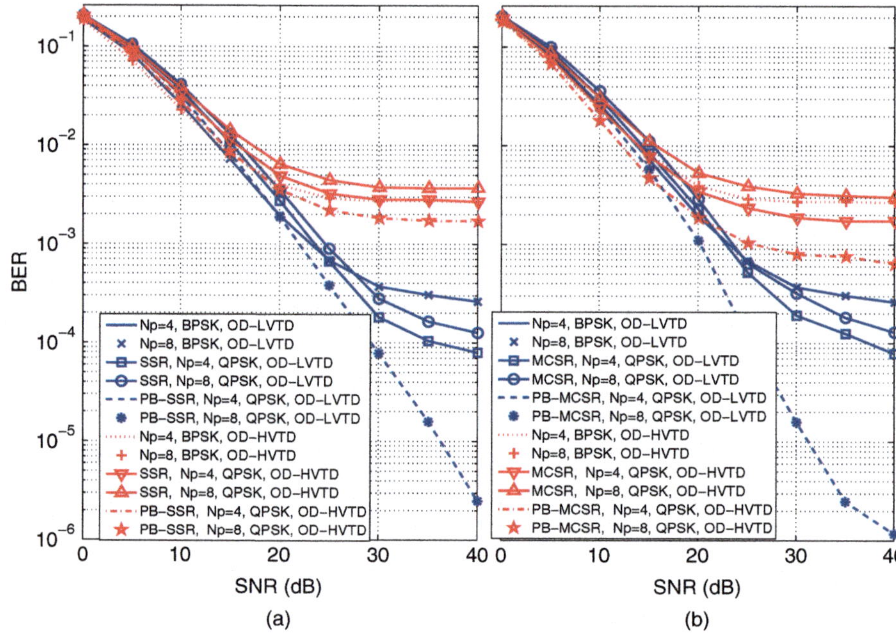

Fig. 3.4 BER performance of (**a**) SSR ICI self-cancellation scheme and PB-SSR scheme and (**b**) MCSR ICI self-cancellation scheme and PB-MCSR scheme with different number of pilots under OD-LVTD and OD-HVTD V2V channels

CPRA and PRA methods becomes wider when the V2V communication system has better OFDM spectral efficiency, i.e., more subcarriers in one OFDM block. This is because that in this case, the weight of Doppler spectrum in the generation of frequency offset becomes larger. Based on the results, it can be clearly obtained that the PBC schemes with the CPRA method are the best ICI cancellation schemes for practical vehicular communication systems.

3.3 Index Modulated (IM-)OFDM

The explosive increase of mobile data services and the use of smart phones require 5G systems to support higher spectrum efficiency (SE), higher energy efficiency (EE), and higher mobility. Conventional MIMO may achieve high SE with massive antennas, but has compromised EE due to the scaled power consumption of a large number of RF chains. The OFDM modulation is prone to Doppler-induced ICI and its inherent high peak-to-average power ratio (PAPR) also necessitates expensive power amplifiers. Novel advanced modulation techniques are therefore needed. To address these concerns, most recently, a new OFDM-based technique, termed as index modulated (IM-)OFDM [15], has arisen as a promising candidate that has the

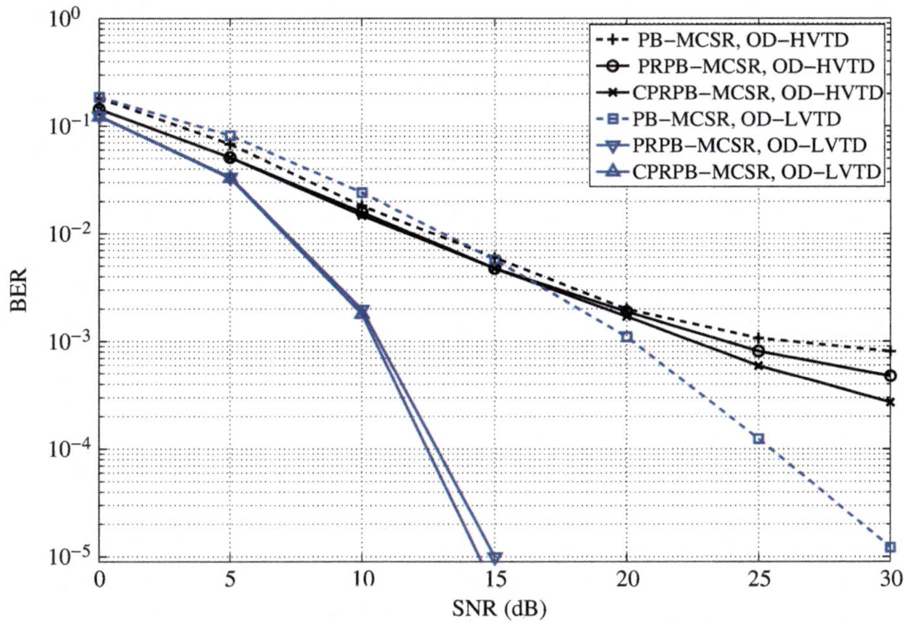

Fig. 3.5 BER performance of PB-MCSR, PRPB-MCSR, and CPRPB-MCSR schemes under OD-LVTD and OD-HVTD V2V channels

potential to meet the 5G requirements. Compared with plain OFDM, IM-OFDM has reduced PAPR, improved EE and BER performance, but with similar complexity [16]. In this section, we will introduce the IM-OFDM concept and provide an effective and feasible IM-OFDM solution for V2X communications, i.e., IM-OFDM with ICI self-cancellation.

3.3.1 IM-OFDM Concept

A typical transceiver structure of IM-OFDM is shown in Fig. 3.6 [17]. Assuming that there are in total N OFDM subcarriers, to ease the implementation of IM-OFDM, the total N subcarriers are split into G sub-blocks, each consisting of $L = N/G$ subcarriers. Then, the subcarrier activation and symbol modulation are performed within each subblocks according to the following procedure independently. Firstly, b out of L subcarriers in each sub-block are set to be active. Then, b symbols are drawn from the M-ary PSK/QAM constellation to be sent from the active subcarriers, where M is the cardinality of the constellation. Finally, the remaining $(L - b)$ are zero-padded. Given L and b, there are in all $\mathbb{C}(L, b)$ combinations of active subcarrier indices, where $\mathbb{C}(\cdot, \cdot)$ denotes the binomial coefficient. Hence, $\lfloor \log_2 \mathbb{C}(L, b) \rfloor$ bits can be modulated to the active subcarrier indices via either a

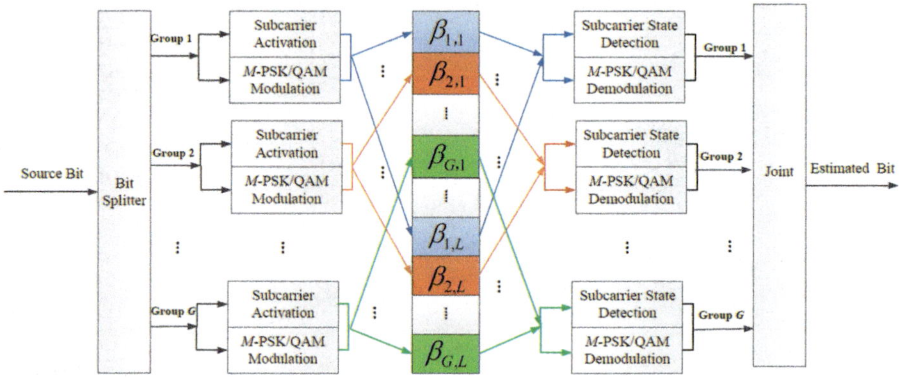

Fig. 3.6 Transceiver Structure of IM-OFDM

look-up table or the combinatorial method proposed in [15], with $\lfloor \cdot \rfloor$ being the floor operation. Therefore, in conjunction with the information bits carried by the b constellation symbols, IM-OFDM can convey a total of $\lfloor \log_2 \mathbb{C}(L, b) \rfloor + b \log_2 M$ bits per sub-block. By concatenating the subcarrier sub-blocks, the original IM-OFDM block can be obtained, which is grouped in a localized manner with its i-th element represented as S_i, $(i = 0, \cdots, N-1)$. To benefit from the uncorrelated channels and improve the system performance, the subcarriers are then go through an interleaving module which results in $\Phi^g = \{g-1, g-1+G, \cdots, g-1+(L-1)G\}$ as depicted in Fig. 3.6, where $\Phi^g = \{\beta_{g,1}, \cdots, \beta_{g,L}\}$, $(g = 1, \cdots, G)$ represents the subcarrier indices of the g-th sub-block, and $\Phi^1 \cup \cdots \cup \Phi^G = \{0, \cdots, N-1\}$. Clearly, the i-th element in localized grouping is placed to the j-th subcarrier after interleaving, with $j = \mathcal{I}(i) = i\%L \times G + \lfloor i/L \rfloor$, $(i = 0, \cdots, N-1)$. Next, the inverse FFT is applied and a cyclic prefix (CP) is added to the beginning of the time-domain OFDM symbol before sent from the transmit antenna.

At the receiver, after CP reduction, FFT and de-interleaving, the received signal within the g-th sub-block in the frequency domain can be expressed by

$$Y_{\beta_{g,l}} = H_{\beta_{g,l}} X_{\beta_{g,l}} + W_{\beta_{g,l}}$$

where $X_{\beta_{g,l}}$ is the transmitted M-PSK/QAM symbol carried on the active subcarrier after interleaving, $H_{\beta_{g,l}}$ is the channel coefficient, and $W_{\beta_{g,l}}$ is the AWGN of variance N_0, at the $\beta_{g,l}$-th subcarrier, respectively. It is worth noting that the average transmit power of $X_{\beta_{g,l}}$ is P/b rather than P/L as in conventional OFDM with P being the total transmit power per sub-block due to the presence of inactive subcarriers.

To demodulate the information bits at the receiver, the log-likelihood ratio (LLR) detector of IM-OFDM evaluates the ratio of the posteriori probability of non-zero to that of zero for each subcarrier. This ratio gives information about the status of the corresponding subcarriers and can be written as

$$\lambda_{\beta_{g,l}} = \ln(b) - \ln(L-b) + \frac{|Y_{\beta_{g,l}}|^2}{N_0} + \ln\left(\sum_{\theta=1}^{M} \exp\left(-\frac{1}{N_0}|Y_{\beta_{g,l}} - H_{\beta_{g,l}} s_\theta|^2\right)\right), \tag{3.10}$$

where $\{s_1, \cdots, s_M\}$ represents the M-ary constellation symbols. The b subcarriers within the g-th sub-block having maximum LLR values are determined to be active. Then the received signals associated with the determined active subcarriers are demodulated to get the estimated information bits. Note that the LLR detector may result in catastrophic error when deciding on an unused index combination. However, since the unused combinations are much less than the used ones, the performance penalty is negligible.

The spectral efficiency of IM-OFDM can be given as [16]

$$\mathrm{SE}_{\mathrm{IM\ OFDM}} = \frac{\log_2 \mathbb{C}\,(L, m)}{L} + \frac{(L-m)\log_2 M}{L} \tag{3.11}$$

where L is the number of subcarriers per group, m is the number of inactive subcarriers, and M is the cardinality of the constellation. Given L and m, there are $\mathbb{C}\,(L, m)$ combinations of active subcarrier indices. Hence, $\log_2 \mathbb{C}\,(L, m)$ bits can be conveyed via subcarrier activation. Therefore, in conjunction with the information bits carried on the $(L-m)$ modulated symbols, IM-OFDM can convey a total of $\left(\log_2 \mathbb{C}\,(L, m) + (L-m)\log_2 M\right)$ bits per sub-block. In [16], the comparison between IM-OFDM and plain OFDM in terms of the maximum achievable rate (MAR) was evaluated, which indicates that with the same data rate, IM-OFDM provides improved MAR compared with plain OFDM and this advantage varies with different m values.

3.3.2 *IM-OFDM with ICI Self-Cancellation*

Since only a subset of subcarriers are active in IM-OFDM, this may also create new opportunities in dealing with ICI in vehicular communications. However, the question lies in that although IM-OFDM is expected to have the potential of ICI suppression, it has been verified that this potential vanishes under severe ICI. This happens because in the presence of high Doppler shift in vehicular communications, the ICI may induce leakage into inactive subcarriers, which increases the possibility of erroneous detection of subcarrier states and in turn misleads the detection of transmitted symbols on the active subcarriers [16]. To solve this problem, ICI self-cancellation was introduced to IM-OFDM to make IM-OFDM feasible and effective in vehicular scenarios, termed as IM-OFDM with ICI self-cancellation [19].

3.3.2.1 Transceiver Design

Figure 3.7 shows the transceiver structure of the IM-OFDM with ICI self-cancellation. From Fig. 3.7, we can see that the pair-wise subcarriers are activated together, as opposed to individual one in ICI cancellation for plain OFDM. The signals with opposite polarity are allocated to the adjacent subcarriers, which makes the ICI signals between adjacent subcarriers self-cancelled.

As for the transmitter structure shown in Fig. 3.7a, for each group, only m out of L subcarrier pairs are set to be active to transmit modulated symbols, while the remaining $L - m$ subcarrier pairs are set to be idle. It is worth noting that the subcarrier state is determined pair-wise in the proposed scheme, which means, the two neighboring subcarriers within one subcarrier pair share the same state. Therefore, for each group, there will be $\mathrm{C}(L, m)$ kinds of combinations which consist of the indices of active subcarrier pairs. Hence, $p_1 = \lfloor \log_2 \mathrm{C}(L, m) \rfloor$ bits can be conveyed by subcarrier activation, where $\mathrm{C}(\cdot, \cdot)$ denotes the binomial coefficient and $\lfloor \cdot \rfloor$ returns the maximal integer less than the argument. In each group, the first coming p_1 bits determine the index combination of active/inactive subcarrier pairs. The determination can be implemented via either the look-up table method or the combinatorial method.

Denote the determined index combination for subcarrier group g as I_g, which is comprised of the indices of active subcarrier pairs. Next, the modulated symbols are drawn from the M-ary PSK/QAM constellation and then assigned to the active subcarrier pairs, where M is the cardinality of the constellation. Due to the fact that the difference between ICI coefficients of adjacent subcarriers is very small, the ICI signal is expected to be self-canceled within adjacent subcarriers. Therefore, a data pair with opposite polarity, such as $(a, -a)$, is modulated onto two adjacent subcarrier $\{\beta_{g,l}^1, \beta_{g,l}^2\}$, where $\beta_{g,l} \in I_g$. With the ICI self-cancellation technique in [10], the ICI generated by $\beta_{g,l}^2$-th subcarrier is expected to be cancelled out by that generated by $\beta_{g,l}^1$-th subcarrier. Since $m \log_2 M$ bits can be conveyed by symbol modulation, by taking subcarrier activation and M-ary modulation into account, each subcarrier group conveys $\lfloor \log_2 \mathrm{C}(L, m) \rfloor + m \log_2 M$ bits. Therefore, the spectral efficiency achieved by IM-OFDM with ICI self-cancellation is (bps/Hz)

$$f_{\text{IM-OFDM}} = \frac{\lfloor \log_2 \mathrm{C}(L, m) \rfloor}{2L} + \frac{m \log_2 M}{2L}. \tag{3.12}$$

Interleaved grouping is also considered for its optimality among all possible subcarrier grouping methods. Different from the manner adopted in conventional IM-OFDM, in the IM-OFDM with ICI self-cancellation scheme, the subcarriers are grouped in pairs. Equal-spaced subcarrier pairs in frequency are collected to form a group

$$\begin{aligned}&\{\beta_{1,1}, \beta_{2,1}, \cdots, \beta_{G,1}, \beta_{1,2}, \beta_{2,2}, \cdots, \beta_{G,2}, \cdots, \beta_{1,L}, \beta_{2,L},\\ &\cdots, \beta_{G,L}\} = \left\{ \bigcup_{g=1}^{G} \beta_{g,1}, \bigcup_{g=1}^{G} \beta_{g,2} \cdots, \bigcup_{g=1}^{G} \beta_{g,L} \right\}.\end{aligned}$$

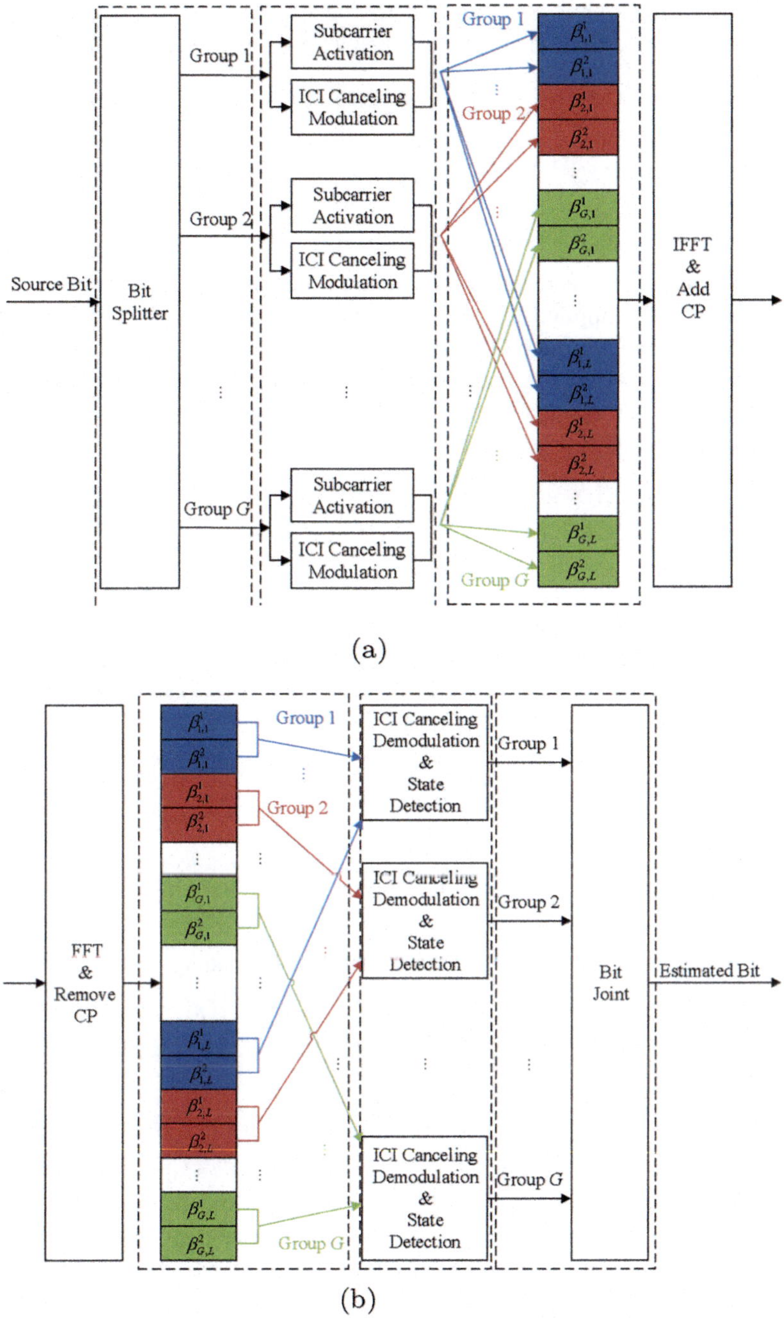

Fig. 3.7 Transceiver structure of IM-OFDM with ICI self-cancellation. (**a**) Transmitter structure. (**b**) Receiver structure

Denote the transmitted signals in the frequency domain as $x(k)$, where $k = 1, 2, \ldots, N$. The transmitted signals at all subcarriers with the IM-OFDM with ICI self-cancellation scheme are given by

$$\mathbf{X} = [x(1), \ldots, x(N)]^T = \left[\bigcup_{g=1}^{G} s_{\beta_{g,1}}, \cdots, \bigcup_{g=1}^{G} s_{\beta_{g,L}} \right]^T, \tag{3.13}$$

where $s_{\beta_{g,l}}$ denotes the $\beta_{g,l}$-th pair of transmitted signals, $s_{\beta_{g,l}} = \{s_{\beta^1_{g,l}}, s_{\beta^2_{g,l}}\}$, and $(\cdot)^T$ stands for the transpose operation. Before transmission, the inverse fast Fourier transform (IFFT) is applied to (3.13). Then, a cyclic prefix (CP) is added to the beginning of the time-domain OFDM symbol.

As for the receiver structure shown in Fig. 3.7b, the CP of the received signal is first removed and the application of the FFT is followed. The received signal at the k-th subcarrier can be expressed as

$$y(k) = h(k)x(k)C(0) + \sum_{i=1, i\neq k}^{N} h(i)x(i)C(i-k) + w(k), \tag{3.14}$$

where $h(k)$ and $w(k)$ represent the channel frequency response and the noise at the k-th subcarrier respectively, where $k = 1, 2, \cdots, N$. The second term at the right-hand side of (3.14) represents the ICI, where $C(i-k)$ is the ICI coefficient, given by

$$\begin{aligned} C(i-k) =& \frac{\sin(\pi(i+\varepsilon-k))}{N \sin\left(\frac{\pi}{N}(i+\varepsilon-k)\right)} \\ & \cdot \exp\left(\frac{j\pi(N-1)(i+\varepsilon-k)}{N}\right), \end{aligned} \tag{3.15}$$

with ε denoting the normalized CFO to the subcarrier spacing.

Denote the $\beta_{g,l}$-th received signal pair as $r_{\beta_{g,l}} = \left\{r_{\beta^1_{g,l}}, r_{\beta^2_{g,l}}\right\}$. The received signals at all subcarriers with the IM-OFDM with ICI self-cancellation scheme can be given by

$$\mathbf{Y} = [y(1), \ldots, y(N)]^T = \left[\bigcup_{g=1}^{G} r_{\beta_{g,1}}, \cdots, \bigcup_{g=1}^{G} r_{\beta_{g,L}} \right]^T. \tag{3.16}$$

Next, all received signals are grouped in an interleaved manner similar to the operations performed at the transmitter. That is, all received signals are extracted as $\left\{r_{\beta_{1,1}}, r_{\beta_{1,2}}, \cdots, r_{\beta_{1,L}}, \cdots\cdots, r_{\beta_{G,1}}, r_{\beta_{G,2}}, \cdots, r_{\beta_{G,L}}\right\} = \{\bigcup_{l=1}^{L} r_{\beta_{1,l}}, \cdots, \bigcup_{l=1}^{L} r_{\beta_{G,l}}\}$. Then, the ICI cancelling demodulation is executed within each pair of subcarriers, yielding

$$r'_{\beta_{g,l}} = \frac{1}{2}(r_{\beta^1_{g,l}} - r_{\beta^2_{g,l}}). \tag{3.17}$$

In this manner, the residual ICI in the received signals can be reduced further.

The joint detection of the subcarriers states and the modulated symbols is carried out based on the maximum-likelihood (ML) criterion. The ML detector considers all possible realizations by searching for all possible subcarrier index combinations and signal constellation points to make a joint decision on the combination of active indices and the constellation symbols, denoted as $\hat{I}_g$ and $\hat{S}_g$ respectively. For group g, there are $\mathrm{C}(L, m)$ kinds of index combinations of active subcarriers, denoted by Φ^g_b, where $b = 1, 2, \cdots, \mathrm{C}(L, m)$. Let $\bar{\Phi}^g_b$ denote the complement of Φ^b_g. The ML detector can be derived as

$$\left\langle \hat{I}_g, \hat{S}_g \right\rangle = \underset{b \in \Theta}{\arg\min} \sum_{\beta_{g,l} \in \Phi^g_b} \left| r'_{\beta_{g,l}} - h'_{\beta_{g,l}} \hat{s}_{\beta_{g,l}} \right|^2 + \sum_{\beta_{g,l} \in \bar{\Phi}^g_b} \left| r'_{\beta_{g,l}} \right|^2, \tag{3.18}$$

where $\Theta = \{1, 2, \cdots, \mathrm{C}(L, m)\}$ and $\hat{s}_{\beta_{g,l}}$ is the estimated transmitted signal carried on the $\beta_{g,l}$-th subcarrier pair, which can be gotten by searching for the closest constellation point to $r'_{\beta_{g,l}}$, and $h'_{\beta_{g,l}}$ is the channel frequency response at the $\beta_{g,l}$-th subcarrier pair, defined as $h'_{\beta_{g,l}} = \frac{1}{2}\left(\hat{h}_{\beta^1_{g,l}} + \hat{h}_{\beta^2_{g,l}}\right)$, where $\hat{h}_{\beta^1_{g,l}}$ and $\hat{h}_{\beta^2_{g,l}}$ represent the channel estimates on the corresponding subcarriers.

3.3.2.2 Simulation Verifications

In order to evaluate the IM-OFDM with ICI self-cancellation scheme, in [18, 19], simulations were carried out by employing the V2X channel model proposed in [20], which contains six scenarios

- Scenario 1: V2V Expressway Oncoming;
- Scenario 2: V2V Expressway Same Direction with Wall;
- Scenario 3: V2V Urban Canyon Oncoming;
- Scenario 4: Roadside-to-vehicle (R2V) Expressway;
- Scenario 5: R2V Urban Canyon Oncoming; and
- Scenario 6: R2V Suburban Street.

Except for the proposed IM-OFDM with ICI self-cancellation scheme, three other schemes are chosen as baselines for performance comparison:

- Scheme 1: OFDM with BPSK modulation;
- Scheme 2: OFDM with ICI self-cancellation and QPSK modulation;
- Scheme 3: IM-OFDM with four subcarriers as a group, two subcarriers activated, and BPSK modulation.

Figures 3.8 and 3.9 show the performance differences among Schemes 1–3 and the proposed IM-OFDM with ICI self-cancellation scheme ($G = 8$, $L = 4$, $m = 3$,

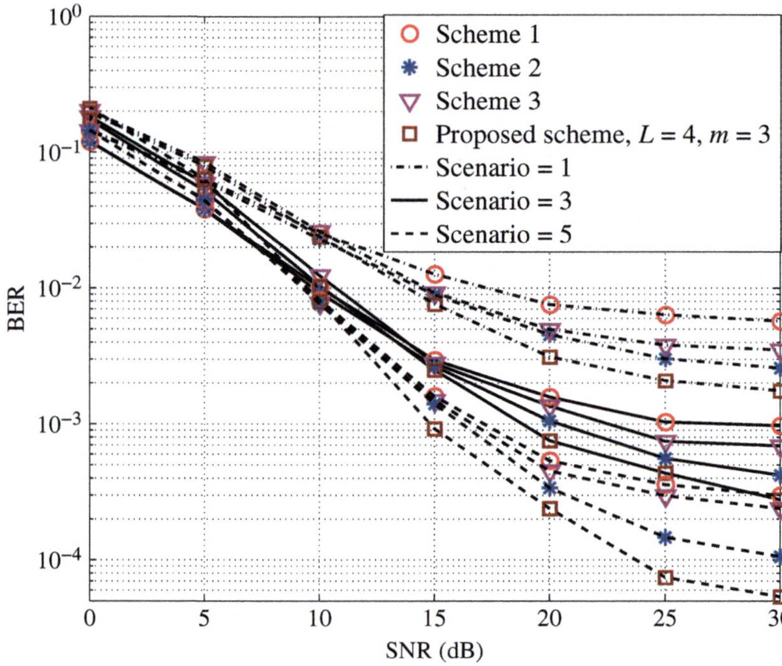

Fig. 3.8 BER performance comparison for Scenario 1, Scenario 3, and Scenario 5 among Schemes 1–3 and the proposed IM-OFDM with ICI self-cancellation scheme ($L = 4$, $m = 3$, and QPSK modulation)

and QPSK modulation) under different scenarios. It can be seen that when SNR is low, Scheme 1 and Scheme 2 perform better than the IM-OFDM and the IM-OFDM with ICI self-cancellation scheme. The reason lies in that when the SNR is low, the noise largely contributes to the power of received signals. Consequently, the power of the received signal at each inactive subcarrier becomes comparable to that at each active subcarrier, which also misleads the detection of subcarrier states. To make the IM-OFDM with ICI self-cancellation scheme more practical, when SNR$<$ 10 dB, all pairs of subcarriers in the proposed scheme can be set as active. In this way, there is no need to perform subcarrier states detection at receiver and the IM-OFDM with ICI self-cancellation scheme is equivalent to the OFDM with ICI self-cancellation scheme, which achieves a better BER performance when SNR is low.

As the SNR increases, it can be found that IM-OFDM and proposed scheme outperform conventional OFDM. This happens because subcarriers in IM-OFDM and the proposed scheme are grouped in an interleaved manner, as a result, the effect of correlated frequency-selective fading in V2X channels is reduced significantly. However, IM-OFDM performs worse than Scheme 2 and exhibits error floor at 25 dB. It can be understood since that the ICI signals caused by Doppler shift largely deteriorate the system performance of IM-OFDM, while the ICI signals

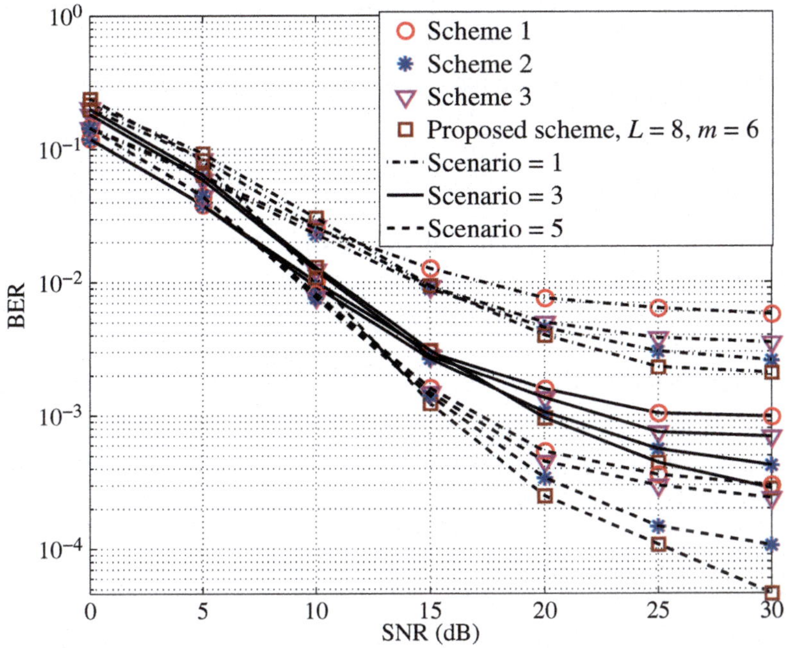

Fig. 3.9 BER performance comparison for Scenario 1, Scenario 3, and Scenario 5 among Schemes 1–3 and the proposed IM-OFDM with ICI self-cancellation scheme ($L = 8$, $m = 6$ and QPSK modulation)

can be suppressed due to the ICI self-cancellation method in Scheme 2. As shown in Fig. 3.8, when BER falls below 10^{-2}, the IM-OFDM with ICI self-cancellation scheme performs better than conventional OFDM and IM-OFDM under different scenarios, and more importantly, even better than OFDM with ICI self-cancellation. This is as expected because in the IM-OFDM with ICI self-cancellation scheme, with the ICI self-cancellation technique being integrated into the IM-OFDM frame work, almost $N(L-m)/L$ subcarriers transmit zero energy and the number of active subcarriers which actually incur ICI is reduced to $(m/L)N$.

In the simulations, Scenario 1 has the highest Doppler shift. Correspondingly, the BER error floors are the highest for all the schemes. However, it is worth mentioning that, the IM-OFDM with ICI self-cancellation results in the most significant performance boost when the Doppler is relatively high. This is because the as the Doppler-induced ICI becomes dominant, the effect of ICI suppression in the proposed scheme becomes more obvious with the adoption of the ICI self-cancellation technique.

3.4 Differential Spatial Modulation

As a rising star in MIMO research, spatial modulation (SM) has attracted significant interests from both academia and industry, since it achieves diversity and multiplexing gains while using only a single RF chain. To cope with the fast time-varying channels and effectively employ this promising MIMO technique in vehicular communications, differential spatial modulation (DSM) was proposed in [21, 22]. In this section, we mainly introduce the DSM technique and its application in V2X communications.

3.4.1 DSM Concept

Spatial modulation [23] is a recently developed MIMO technique, in which the information bits are partitioned into two parts, of which one modulates the symbol and the other activates one out of all transmit antennas. The property of SM that only a single antenna is activated for transmission allows SM to eliminate ICI and to avoid synchronization among the transmit antennas while achieving a spatial multiplexing gain. Due to its potential of realizing low-complexity and spectrally-efficient MIMO implementations, SM has motivated considerable researches. In order to release some limitations of SM, a differential scheme specifically tailored to SM, namely DSM, was proposed in [21], which also exploits the time dimension in addition to space to facilitate the differential (de-)modulation.

3.4.1.1 From SM to DSM

Consider an SM system with N_t transmit antennas and N_r receive antennas that operates over a Rayleigh flat-fading channel. Assume that at time instant t, symbol s_t is transmitted via the mth transmit antenna. The constellation vector can be written as $\mathbf{s} = \left[0 \cdots 0 \; s_t \; 0 \cdots 0\right]^T$, where the only nonzero entry is the mth element of the N_t-dimensioned column vector. The received signal at the SM receiver then can be expressed as $\mathbf{y} = \mathbf{h}_m s_t + \mathbf{n}$, where $\mathbf{h}_m$ collects the N_r channels from the mth transmit antenna to the N_r receive antennas and $\mathbf{n}$ is an N_r-dimensioned additive white Gaussian noise (AWGN) column vector at the N_r receive antennas with covariance matrix $N_0\mathbf{I}_{N_r}$. SM exploits the index of the activated transmit antenna as an additional information conveying mechanism. Therefore, the transmit-to-receive wireless link is time hopping in accordance with the shifting activated transmit antenna. This hopping feature of SM transmission renders the development of its differential scheme a challenging issue.

In the DSM, the original SM constellation vector $\mathbf{s}$ in time is collected to form an $N_t \times N_t$ space-time block $\mathbf{S}$. The (m, t)-th entry of $\mathbf{S}$ denotes the symbol s_{mt}

transmitted via transmit antenna m at time instant t. The $N_t \times N_t$ block $\mathbf{S}$ is required to satisfy the following conditions:

(1) Only one antenna remains active at each time instant, that is, only one entry in any column of $\mathbf{S}$ is nonzero.
(2) Each antenna is activated once and only once in the N_t successive time instants of a space-time block, that is, only one entry in any row of $\mathbf{S}$ is nonzero.
(3) The signal constellation is restricted to an equal energy 2^b-PSK alphabet $\mathcal{A}$ for some $b = 1, 2, 3 \cdots$, that is, each nonzero entry is chosen from $\mathcal{A}$. A special case of $\mathcal{A}$ with $b = 0$ can be referred to as differential space shift keying (DSSK), in which all the transmitted symbols are 1 s.

An example of the space-time block for $N_t = 3$ is given by $\mathbf{S} = \begin{bmatrix} s_{11} & 0 & 0 \\ 0 & 0 & s_{23} \\ 0 & s_{32} & 0 \end{bmatrix}$ which means that at time instants 1, 2, and 3, the symbols s_{11}, s_{32}, and s_{23} drawn from $\mathcal{A}$ are sent from transmit antennas 1, 3, and 2, respectively, whereas the other two transmit antennas remain idle. Note that in the DSM, we proposed to exploit the time domain by transmitting blocks of signals using an $N_t \times N_t$ transmission matrix. By this design, each antenna is activated once and only once during each block, thus making differential operation possible so long as the wireless channel remains unchanged over two consecutive blocks. In order to facilitate differential (de-)modulation, one also has to make sure that the transmitted signal blocks satisfy the so-termed closure property (see e.g., [24, Sec.5.2.8] and [25]).

3.4.1.2 Spectral Efficiency of DSM

Let the set $\mathcal{G}$ contain all possible space-time blocks $\mathbf{S}$. Then, the set $\mathcal{G}$ with N_t transmit antennas and 2^b-PSK constellation contains $N_t!2^{N_t b}$ blocks. In theory, the spectral efficiency of the proposed DSM is (bits/s/Hz)

$$R_{\text{DSM, theory}} = \frac{1}{N_t}\log_2(N_t!) + b. \tag{3.19}$$

Based on Stirling's Formula $n! \approx \sqrt{2\pi n}n^n e^{-n}$ [26], the spectral efficiency of DSM can be approximated as

$$\begin{aligned} &R_{\text{DSM, theory}} \\ &\approx \log_2(N_t) - \underbrace{\left(\log_2(e) - \frac{1}{N_t}\log_2\left(\sqrt{2\pi N_t}\right)\right)}_{loss\ of\ spectral\ efficiency} + b \end{aligned} \tag{3.20}$$

where e is the mathematical constant. Note that the spectral efficiency of a conventional SM system is

$$R_{\text{SM, theory}} = \log_2(N_t) + b. \tag{3.21}$$

Therefore, the loss of spectral efficiency of DSM compared to that of SM can be approximated as

$$\log_2(e) - \frac{1}{N_t}\log_2\left(\sqrt{2\pi N_t}\right) \leq \log_2(e). \tag{3.22}$$

This indicates that the loss of the spectral efficiency is upper bounded by $\log_2(e)$.

Note that when $N_t \geq 3$, $N_t!$ is not an integer power of two. One way to approach the theoretical spectral efficiency $R_{\text{DSM, theory}}$ is to apply the so-called Fractional Bit Encoding (FBE) to DSM, which was applied to SM in [27] for the case in which N_t is not an integer power of two. However, FBE leads to longer decoding delay and worse bit error rate (BER) performance.

In practical use of SM, N_t is usually an integer power of 2. Hence the spectral efficiency of SM is

$$\overline{R}_{\text{SM}} = \lfloor \log_2(N_t) \rfloor + b. \tag{3.23}$$

Similarly, a simpler DSM without fractional encoding is preferable. For this purpose, choose $2^{\lfloor \log_2(N_t!) \rfloor} 2^{N_t b}$ transmit blocks from $\mathcal{G}$ and form a subset $\mathcal{G}_M$ for (de-)mapping. Each transmit block can then be encoded with $\lfloor \log_2(N_t!) \rfloor + N_t b$ bits. Therefore, the spectral efficiency of DSM for practical use can be written as

$$\overline{R}_{\text{DSM}} = \frac{1}{N_t} \lfloor \log_2(N_t!) \rfloor + b. \tag{3.24}$$

Figure 3.10 compares the BER performance between DSM and SM using $N_t = 4$ transmit antennas. The targeted spectral efficiency is 3 bits/s/Hz. Note that appropriate modulation orders are employed to obtain the same spectral efficiency, i.e., QPSK for DSM and BPSK for SM. It can be observed that the loss in BER performance induced by the differential scheme is no more than 3 dB. This result indicates that though DSM uses higher modulation order than SM in order to make up for its spectral efficiency loss in the spatial domain, a 3 dB degradation is still retained.

3.4.2 *DSM Transceiver Design*

3.4.2.1 Differential Transmission Process

As illustrated in Fig. 3.11, the transmitter begins the transmission by sending an arbitrary initial block $\mathbf{S}_0 \in \mathcal{G}$. Since the 2^b-PSK constellation $\mathcal{A}$ always contains symbol 1, we choose $\mathbf{S}_0 = \mathbf{I}_{N_t}$ for any N_t without loss of generality. Thereafter,

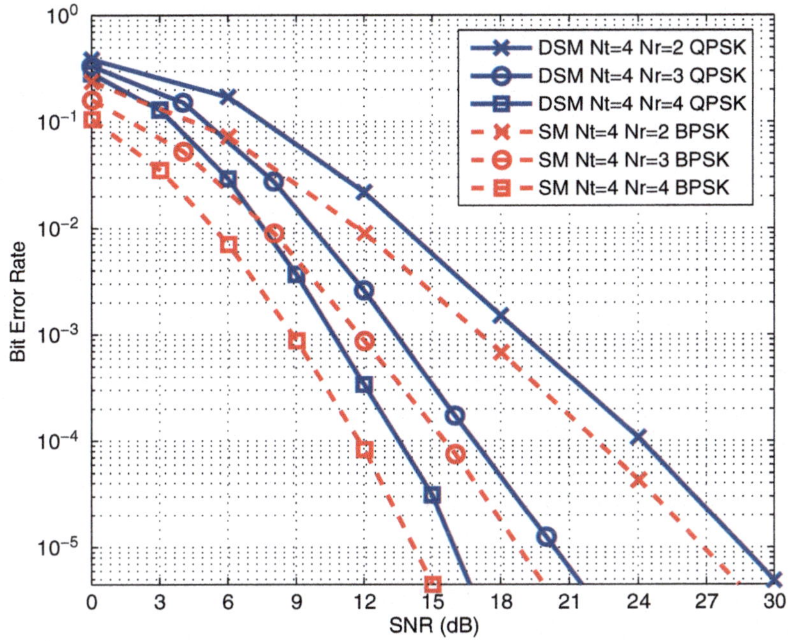

Fig. 3.10 BER performance of differential detection DSM versus coherent detection SM at 3 bits/s/Hz transmission rate with $N_t = 4$. DSM uses QPSK whereas SM uses BPSK

the transmitter encodes the bit stream in a recursive manner. Suppose that from time instant $N_t(t-1)$ to $N_t t - 1$, the $(t-1)$-th block $\mathbf{S}_{t-1}$ is transmitted ($t = 1, 2, 3, \cdots$). This encoding process in the next space-time block operates as follows:

(1) *Step 1*: Given the input bit stream, we map $\lfloor \log_2(N_t!) \rfloor + N_t b$ bits onto $\mathcal{G}_M$ and obtain $\mathbf{X}_t \in \mathcal{G}_M \subseteq \mathcal{G}$.

Without loss of generality, the following mapping is employed. The first $\lfloor \log_2(N_t!) \rfloor$ bits are mapped onto the spatial domain, determining the order of transmit antennas being activated in a space-time block by an index-mapping procedure. Then, the remaining $N_t b$ bits are mapped to the signal domain. Specifically, each group of b bits is mapped to a symbol drawn from the 2^b-PSK constellation $\mathcal{A}$ for all N_t time instants. Taking into account the mapping results of both parts, we can determine $\mathbf{X}_t$.

(2) *Step 2*: Compute the transmitted block $\mathbf{S}_t$ as follows:

$$\mathbf{S}_t = \mathbf{S}_{t-1}\mathbf{X}_t. \tag{3.25}$$

Note that given $\mathbf{S}_{t-1}$ and $\mathbf{X}_t \in \mathcal{G}$, according to the so-termed closure property, it is guaranteed that $\mathbf{S}_t \in \mathcal{G}$.

(3) *Step 3*: Send the block $\mathbf{S}_t$ during time period $N_t t$ to $N_t(t+1) - 1$.

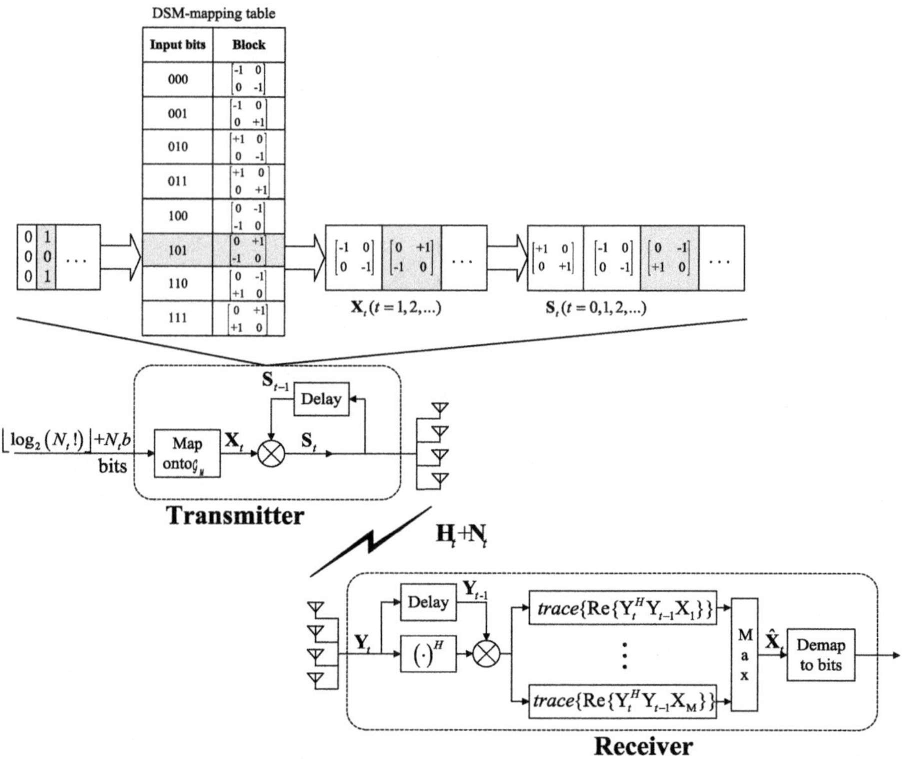

Fig. 3.11 Illustrated DSM transmitter and receiver design. An example of the mapping table to the space-time blocks in the case of $N_t = 2$ BPSK is depicted

The transmitter repeats the above three-step process till the end of the transmission.

3.4.2.2 Differential Detection

Suppose that the $(t-1)$-th and the t-th received signal blocks are $\mathbf{Y}_{t-1}$ and $\mathbf{Y}_t$, respectively. Then, we have

$$\mathbf{Y}_{t-1} = \mathbf{H}_{t-1}\mathbf{S}_{t-1} + \mathbf{N}_{t-1} \tag{3.26}$$

and

$$\mathbf{Y}_t = \mathbf{H}_t\mathbf{S}_t + \mathbf{N}_t. \tag{3.27}$$

Assume a quasi-static channel in which the fading coefficients remain constant over two adjacent DSM transmit blocks, namely $2N_t$ symbol durations. By substituting (3.25) and (3.26) into (3.27), we have

$$\mathbf{Y}_t = \mathbf{Y}_{t-1}\mathbf{X}_t - \mathbf{N}_{t-1}\mathbf{X}_t + \mathbf{N}_t. \tag{3.28}$$

Therefore, the optimal maximum-likelihood (ML) detector can be derived as

$$\hat{\mathbf{X}}_t = \arg\min_{\forall \mathbf{X} \in \mathcal{G}_M} \parallel \mathbf{Y}_t - \mathbf{Y}_{t-1}\mathbf{X} \parallel_F^2 . \tag{3.29}$$

Applying the identity that $trace\{\mathbf{AB}\} = trace\{\mathbf{BA}\}$, (3.29) can be reduced to

$$\begin{aligned} \hat{\mathbf{X}}_t &= \arg\min_{\forall \mathbf{X} \in \mathcal{G}_M} trace\left\{(\mathbf{Y}_t - \mathbf{Y}_{t-1}\mathbf{X})^H (\mathbf{Y}_t - \mathbf{Y}_{t-1}\mathbf{X})\right\} \\ &= \arg\max_{\forall \mathbf{X} \in \mathcal{G}_M} trace\left\{\mathrm{Re}\left\{\mathbf{Y}_t^H \mathbf{Y}_{t-1}\mathbf{X}\right\}\right\}. \end{aligned} \tag{3.30}$$

3.4.2.3 Implementation of The Index-Mapping

The index-mapping procedure maps the incoming bits to the order of transmit antennas being activated in a space-time block, which can be denoted by the permutation of N_t indices of transmit antennas. In the DSM transceiver, the index-mapping procedures are presented in the following:

(1) *Look-up Table Method:*
In this method, a look-up table is created to provide the corresponding permutations for the incoming $\lfloor \log_2(N_t!) \rfloor$ bits. An example for $N_t = 3$ is presented in Table 3.1. Since $3! = 6$, two permutations out of the six are discarded. The selection of permutations may affect the overall performance. For ease of implementation of permutation index-mapping, we choose to discard the lexicographically larger permutations in this development. For arbitrary N_t, the size of the look-up table is $2^{\lfloor \log_2(N_t!) \rfloor}$. This is an efficient and simple method

Table 3.1 An example of look-up table for $N_t = 3$

Bits	Permutation of indices	Blocks
00	(1 2 3)	$\begin{bmatrix} s_{11} & 0 & 0 \\ 0 & s_{22} & 0 \\ 0 & 0 & s_{33} \end{bmatrix}$
01	(1 3 2)	$\begin{bmatrix} s_{11} & 0 & 0 \\ 0 & 0 & s_{23} \\ 0 & s_{32} & 0 \end{bmatrix}$
10	(2 1 3)	$\begin{bmatrix} 0 & s_{12} & 0 \\ s_{21} & 0 & 0 \\ 0 & 0 & s_{33} \end{bmatrix}$
11	(2 3 1)	$\begin{bmatrix} 0 & 0 & s_{13} \\ s_{21} & 0 & 0 \\ 0 & s_{32} & 0 \end{bmatrix}$

when the size of the table is small. However, it is not feasible for larger values of N_t as the table size grows exponentially.

(2) *Permutation Method:* This method introduces a one-to-one mapping between integers and permutations of N_t elements in lexicographical order, i.e., it maps an integer m to a sequence $\mathbf{a}^{(m)} = (a_1^{(m)}, \cdots, a_{N_t}^{(m)})$ which is a permutation of the set $\{1, \cdots, N_t\}$. For fixed N_t, all $m \in [0, N_t! - 1]$ can be presented by a permutation $\mathbf{a}^{(m)}$ of length N_t. This mapping is the so-called Lehmer code [44]. Application of this method to DSM is presented as follows

(i) *Convert the integer m to its factorial representation $\mathbf{b}^{(m)}$ of length N_t.* We define the factorial sequence $\mathbf{b}^{(m)} = (b_1^{(m)}, \cdots, b_{N_t}^{(m)})$, which takes elements according to the following equation

$$m = b_1^{(m)}(N_t - 1)! + \cdots + b_{N_t}^{(m)} 0!. \tag{3.31}$$

As an example, for $N_t = 4$ and $m = 11$, the following factorial sequence $\mathbf{b}^{(m)}$ can be calculated

$$\begin{aligned} &11 = 1 \cdot 3! + 2 \cdot 2! + 1 \cdot 1! + 0 \cdot 0! \\ &\rightarrow \mathbf{b}^{(11)} = (1, 2, 1, 0). \end{aligned} \tag{3.32}$$

The algorithm, which calculates the factorial sequence $\mathbf{b}^{(m)}$, starts by choosing the maximal $b_1^{(m)}$ that satisfies $b_1^{(m)}(N_t - 1)! \leq m$ and chooses the maximal $b_2^{(m)}$ that satisfies $b_2^{(m)}(N_t - 2)! \leq m - b_1^{(m)}(N_t - 1)!$, and so on.

(ii) *Map the factorial sequence $\mathbf{b}^{(m)}$ to a permutation $\mathbf{a}^{(m)}$.* We define $\Theta = (1, 2, \cdots, N_t)$ as an ordered list with its first element zero indexed. The leftmost permutation element $a_1^{(m)}$ is chosen as $\Theta_{b_1^{(m)}}$ and then the element $\Theta_{b_1^{(m)}}$ is removed from the list $\Theta = (1, 2, \cdots, N_t)$. Think of this new list as zero indexed and one can obtain each successive element of $\mathbf{a}^{(m)}$ in a recursive manner:

For $l = 1 : N_t$

$a_l^{(m)} = \Theta_{b_l^{(m)}}$

$\Theta_{b_l^{(m)}}$ is removed from Θ

End For

As an example with $N_t = 4$ and $m = 11$, from the above procedure we have $\mathbf{b}^{(11)} = (1, 2, 1, 0)$. Letting $\Theta = (1, 2, 3, 4)$, we have

$$a_1^{(m)} = \Theta_{b_1^{(m)}} = \Theta_1 = 2 \to \Theta = (1, 3, 4)$$

$$\to a_2^{(m)} = \Theta_{b_2^{(m)}} = \Theta_2 = 4 \to \Theta = (1, 3)$$

$$\to a_3^{(m)} = \Theta_{b_3^{(m)}} = \Theta_1 = 3 \to \Theta = (1)$$

$$\to a_4^{(m)} = \Theta_{b_4^{(m)}} = \Theta_0 = 1 \to \Theta = ()$$

Hence, we obtain $\mathbf{a}^{(m)} = (2, 4, 3, 1)$.

In our proposed scheme in [21], for each block, we first convert the incoming $\lfloor \log_2(N_t!) \rfloor$ bits to an integer m, and then map m to $\mathbf{a}^{(m)}$, dictating the activation order of transmit antennas. At the receiver side, the demapping procedure is straightforward. We can reverse the mapping process and get m from $\mathbf{a}$. Then the integer m is converted to the $\lfloor \log_2(N_t!) \rfloor$ bits.

3.4.3 DSM in V2X Communications

The BER performance of DSM and SM in some typical vehicular scenarios (e.g., V2V expressway oncoming and V2V expressway same direction with wall in [20]) were evaluated in [22]. As shown in Fig. 3.12, due to the Doppler-effect, all the curves exhibit BER floor at high SNR region. Since Scenario 1 has a higher doppler shift, the BER error floors under Scenario 1 for both SM and DSM are higher than those under Scenario 3. For SM system at the same data rate, due to the doppler-effect-introduced ICI, 4-transmit-antenna system performs poorer than 2-transmit-antenna system in spite of the increase in preambles. Nevertheless, as the number of receive antennas increases, the performance gap narrows at high SNR, and they finally reach the same BER floor when $N_r = 2$. Results shown in Fig. 3.12 indicate that not only that DSM outperforms conventional SM, but also the benefits even increases as the Doppler increases. Therefore, DSM is a promising MIMO technique that can be adopted in vehicular communications.

3.5 Energy Harvesting (EH)-Based Vehicular Communications

In vehicular networks, there is a trend to exploit distributed and cooperative MIMO to improve the reliability and efficiency of vehicular networks. Nevertheless, how to make the vehicles willing to participate in the cooperation remains a practical problem. Energy harvesting (EH)-based vehicular communications provides a possible solution, in which the vehicles that help others to perform cooperative

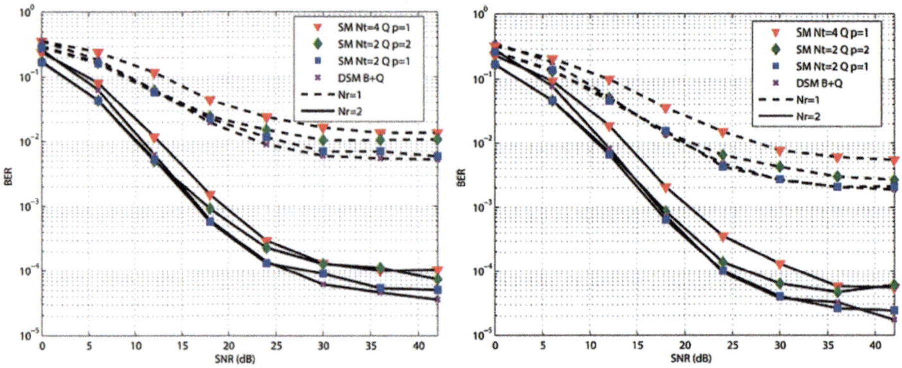

(a) Scenario 1 (2bps/Hz) (b) Scenario 3 (2bps/Hz)

Fig. 3.12 BER performance comparisons for DSM and SM. Scenario 1: V2V Expressway Oncoming; Scenario 3: V2V Expressway Same Direction with Wall.

MIMO can be powered by the received signals without consuming their own energy. Simultaneous wireless information and power transfer (SWIPT) is a promising technique to realize power and information delivery at the same time in wireless communication systems [28]. In this section, we discuss the applications of SWIPT and EH relaying in vehicular networks.

3.5.1 SWIPT over Doubly-Selective Channels

In recent years, SWIPT has drawn increasing interest in the field of wireless communications, signal processing, and networking because of its potential to extend the lifetime of energy constraint systems [29–31]. With SWIPT, useful information and electric power can be transferred by the same radio-frequency (RF) signals simultaneously. Vehicular communications is also a typical scenario in which SWIPT can facilitate the communications of energy constrained sensors. For example, direct connection to the power grid may be unavailable for roadside units (RSUs) in rural areas or along remote highways [32]. These RSUs rely on batteries to power themselves. SWIPT is a viable and affordable solution to extend the lifetime of these RSUs and reduce their maintenance cost. For future vehicles with a rapidly growing number of on-board sensors, SWIPT is more meaningful for sustainable communications of their on-board sensors, whose power can be supplied from other vehicles with excess power or grid-connected RSUs. And this is especially true for electric vehicles as their power is usually more stringent and precious. Moreover, this is also relevant in cooperative communications, where the relay node consumes its own energy to help enhance the communication performance among other nodes. The wireless energy transferred via SWIPT can serve as an incentive for cooperation.

In [33], SWIPT was in vehicular communications and the joint optimization of power allocation and splitting for SWIPT over doubly-selective channels was investigated. Joint power allocation and splitting (JoPAS) and decoupled power allocation and splitting (DePAS) methods were further proposed to achieve near-optimal performance with low complexity.

3.5.1.1 Power Allocation and Splitting Optimization Problem

By taking doubly-selective vehicular channels into consideration, the SWIPT-based vehicular communications is formulated as a joint optimization problem of power allocation, across time and frequency, and SWIPT PS factors, across time slots, in order to maximize the transmission rate for information decoding (ID) within a window of N time slots and K sub-bands. The channel response at the ith subcarrier in the jth time slot is denoted by h_{ij}. Perfect CSI is assumed at the transmitter. Note that although the channel responses are different across both time and frequency, the PS factor can only vary across time, as the PS is implemented at the RF band before any digital processing.

Then, the power allocation and splitting optimization problem can be given

$$\underset{\boldsymbol{P},\boldsymbol{\rho}}{\text{maximize}} \quad \sum_{j=0}^{N-1}\sum_{i=0}^{K-1} \log_2\left(1+\frac{(1-\rho_j)|h_{ij}|^2 P_{ij}}{\sigma^2}\right) \tag{3.33a}$$

$$\text{subject to} \quad \sum_{j=0}^{N-1}\sum_{i=0}^{K-1} P_{ij} \leqslant E_{\text{total}}, \tag{3.33b}$$

$$\rho_k \in [0,1], \tag{3.33c}$$

$$\sum_{j=0}^{N-1} \rho_j \sum_{i=0}^{K-1} h_{ij} P_{ij} \geqslant E_{\text{del}}, \tag{3.33d}$$

$$0 \leqslant \sum_{i=0}^{K-1} P_{ij} \leqslant P_{\max}, \tag{3.33e}$$

$$j = 0, 1, \cdots, N-1, \tag{3.33f}$$

where P_{ij} denotes the transmit power for subcarrier i at time slot j and σ^2 is the variance of the additive white Gaussian noise generated by the down conversion circuits at the receiver. The symbol $\boldsymbol{P}$ represents the vector whose elements are $P_1, P_2, \cdots, P_N$. Without loss of generality, the energy harvest efficiency is assumed to be 1 for simplicity of the analysis.

The constraint (3.33b) limits the total energy consumed by the transmitter during the N-slot time window. It can also be deemed as a limit on the average power consumption at the transmitter because the transmitter could also be energy-constrained. The variable ρ_j represents the portion of power split to energy harvesting at the receiver in the jth time slot and (3.33c) ensures it falls in the range from 0 to 1. The constraint (3.33d) states that the total delivered power to the receiver must exceed P_{del}. The constraints in (3.33e) describe the hardware limitations of instantaneous transmit power for each time slot.

3.5.1.2 Joint Power Allocation and Splitting (JoPAS)

Since the objective function of the maximization problem in (3.33) is non-concave, it is a non-convex optimization problem and thus is difficult to solve efficiently. A two-stage solution to this problem, namely joint power allocation and splitting (JoPAS) is then proposed, as provided in Algorithm 1.

In the first stage, the power allocation and the power splitting factors across the N time slots are jointly optimized. The frequency selectiveness of the channels is ignored and the average channel response is used in the optimization. More specifically, the average channel gain in the jth time slot is given as $\eta_j = \frac{1}{K}\sum_{i=0}^{K-1}|h_{ij}|^2$ and the channel is regarded as flat-fading. By substituting each cross term $(1-\rho_j)P_j$ with an auxilliary variable Q_j, the reformulated JoPAS problem across time can be presented as

$$\underset{\boldsymbol{P},\boldsymbol{Q}}{\text{minimize}} \quad -\sum_{j=0}^{N-1}\ln(K\sigma^2+\eta_j Q_j) \tag{3.34a}$$

$$\text{subject to} \quad \sum_{j=0}^{N-1} P_j \leqslant E_{\text{total}}, \tag{3.34b}$$

$$\sum_{j=0}^{N-1}\eta_j\left(P_j-Q_j\right) \geqslant E'_{\text{del}}, \tag{3.34c}$$

$$P_k \leqslant P_{\max}, \tag{3.34d}$$

$$0 \leqslant Q_k \leqslant P_k, \tag{3.34e}$$

$$k = 0, 1, \cdots, N-1, \tag{3.34f}$$

and ρ_j^* can then be obtained by $\rho_j^* = 1 - \frac{Q_j^*}{P_j^*}$, where $P_j^* \neq 0$. ρ_j^* is set to zero if $P_j^* = 0$. After the reformulation, (3.34) can be solved in polynomial time by many existing algorithms, such as the interior point method [34].

In the second stage, the equivalent power allocated for ID in each time slot $(1-\rho_j)P_j$ is further distributed to each subcarrier by employing the water-filling algorithm, which is known for the optimal power allocation for information transmission over frequency selective channels. Finally, the transmit power for

Algorithm 1 JoPAS

Initialize $E'_{\text{del}} \leftarrow E_{\text{del}}$;
repeat
  Solve (3.34) and obtain the optimizers $\boldsymbol{P}^*$ and $\boldsymbol{Q}^*$.
  for all $j = 1, 2, \cdots, N$ **do**
    if $P_j^* > 0$ **then**
      $\rho_j^* \leftarrow 1 - \frac{Q_j^*}{P_j^*}$.
    else
      $\rho_j^* \leftarrow 0$.
    end if
    if $Q_j^* > 0$ **then**
      Obtain Q_{ij}^* by distributing Q_j^* to all subcarriers with the water-filling algorithm
      $P_{ij}^* \leftarrow Q_{ij}^*/(1 - \rho_j^*)$
    else
      Allocate P_j^* to the subcarrier with the largest $|h_{ij}|^2$.
    end if
  end for
  Compute $\tilde{E}_{\text{del}} \leftarrow \sum_{j=0}^{N-1} \rho_j^* \sum_{i=0}^{K-1} |h_{ij}|^2 P_{ij}^*$.
  if $\tilde{E}_{\text{del}} < E_{\text{del}}$ **then**
    Increase E'_{del}
  else
    Decrease E'_{del}
  end if
until The delivered energy $\tilde{E}_{\text{del}}$ matches E_{del}
Calculate $\rho_k = 1 - \frac{Q_k}{P_k}$.

each subcarrier is obtained by scaling the equivalent power for ID by a factor of $\frac{1}{1-\rho_j}$. Obviously, the optimal solution of (3.34) should satisfy the equality in constraint (3.34c). If the power is evenly distributed to all the subcarriers in a time slot, the delivered energy should also equal to E'_{del}. However, when the transmit power is distributed to each subcarrier according to water-filling, the actual delivered energy will be larger, since the water-filling algorithm allocates higher power to subcarriers with larger channel gains. Therefore, we need to adjust the intermediate energy delivery objective E'_{del} so that the actual delivered energy matches E_{del}.

3.5.1.3 Decoupled Power Allocation and Splitting (DePAS)

Although the first stage of the JoPAS algorithm is a convex optimization problem, the complexity to find the optimum solution is still polynomial with respect to the number of time slots N, which is still a bit high in practical vehicular communications with fast time-variant channels. Whereas the complexity of the water-filling algorithm for traditional power allocation problems is only linear to N. On the other hand, according to Proposition 2 proved in [33], the power allocation for ID follows a pattern similar to water-filling except at time slots

Algorithm 2 DePAS

Initialize $\rho \leftarrow \rho_0$;
repeat
  Split $(1-\rho)P_{\text{total}}$ for ID and allocate Q_j by the waterfilling algorithm
  for all $Q_j > P_{\max}$ **do**
    $Q_j \leftarrow P_{\max}$
  end for
  Prioritize the time slots with larger η_j-s when allocating the remaining power for EH as long as the constraints $P_j \leqslant P_{\max}$ hold.
  for all $j = 1, 2, \cdots, N$ **do**
    if $P_j > 0$ **then**
      $\rho_j \leftarrow 1 - \frac{Q_j}{P_j}$.
    else
      $\rho_j \leftarrow 0$.
    end if
    if $Q_j > 0$ **then**
      Obtain Q_{ij} by distributing Q_j to all subcarriers with the waterfilling algorithm
      $P_{ij} \leftarrow Q_{ij}/(1-\rho_j)$
    else
      Allocate P_j to the subcarrier with the largest $|h_{ij}|^2$.
    end if
  end for
  Compute $\tilde{E}_{\text{del}} \leftarrow \sum_{j=0}^{N-1} \rho_j^* \sum_{i=0}^{K-1} |h_{ij}|^2 P_{ij}^*$.
  if $\tilde{E}_{\text{del}} < E_{\text{del}}$ **then**
    Increase ρ
  else
    Decrease ρ
  end if
until The delivered energy $\tilde{E}_{\text{del}}$ matches E_{del}

where $P_k = P_{\max}$. Hence, a heuristic low-complexity algorithm, namely decoupled power allocation and splitting (DePAS), is further proposed to obtain a sub-optimal solution to the problem in (3.34).

In the DePAS algorithm, as detailed in Algorithm 2, we first split the total transmit power in the N time slots according to a factor ρ. For the part of power split for ID, water-filling scheme is adopted to calculate the power for ID at each time slot. For the other part of power split for EH, time slots with higher η_k are adopted with priority. If the transmit power at the best time slot reaches $P_{\max}$, the channel with the next largest η_k is used for the remaining power and so on. Then, the value of the overall power splitting factor ρ is adjusted until the required power delivery P_{del} is achieved. With similar achievable performance, DePAS can dramatically reduce the computational complexity from $\mathcal{O}(N^3)$ to $\mathcal{O}(N)$.

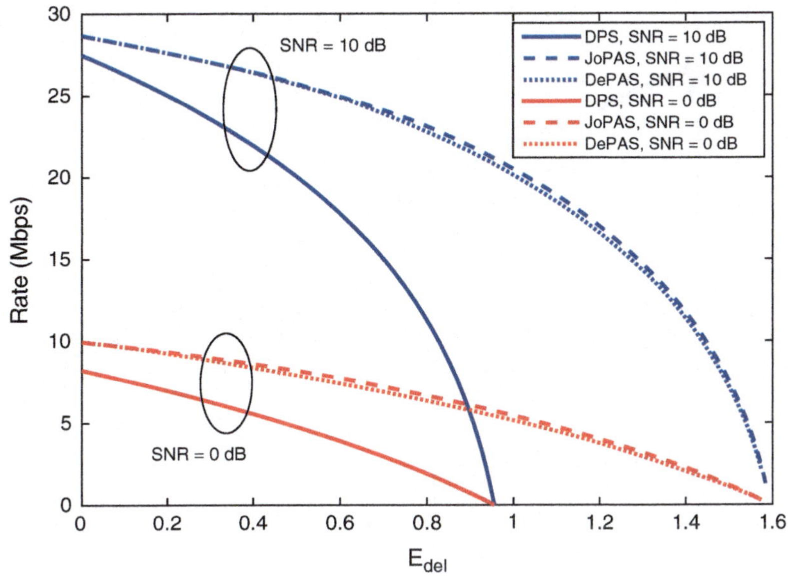

Fig. 3.13 Achievable rate-energy regions of JoPAS, DePAS, and DPS

3.5.1.4 Performance Evaluation

In [33], simulations under doubly-selective vehicular channels were conducted to verify the efficiency of the proposed JoPAS and DePAS. The dynamic power splitting (DPS) [28], which allocates transmit power evenly and optimizes the splitting factor ρ_k at each time slot, is employed as a baseline for performance comparison.

Figure 3.13 shows that both DePAS and JoPAS achieves much larger R-E region than DPS, and the R-E region difference between DePAS and JoPAS is negligible.

The achievable average rate performance comparison is presented in Fig. 3.14. It can be observed that compared with DPS, both JoPAS and DePAS improve the average rate for ID while satisfying the same requirement on EH, especially when larger amount of EH is required. This is because that the DPS only utilizes the CSI for power splitting factor optimization and the power allocation is not adapted according to the CSI. Additionally, the achievable average rate of DePAS is close to that of JoPAS but with much lower complexity.

3.5.2 *EH Relaying in Vehicular Networks*

The integration of SWIPT and cooperative relaying gives rise to EH relaying, which has the potential to further the quest of green 5G communications. 5G

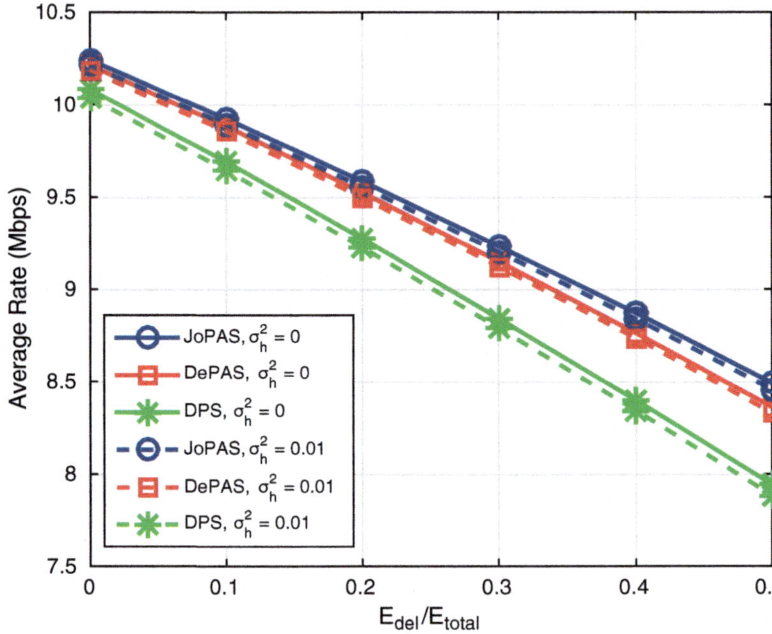

Fig. 3.14 Average rate (Mbps) achieved by JoPAS and DePAS in comparison with DPS under frequency-selective channels. Solid lines represent cases with perfect channel prediction. Dashed lines represent cases with Gaussian channel prediction error with variance 0.01

wireless networks are envisioned to accommodate high mobility user equipments (UEs), which are onboard cars, trucks, or high-speed trains etc. The high speed of UEs results in significant variations and degradations in wireless communication channels. One of the techniques to mitigate such variations is cooperative relaying. It exploits the fact that different UEs experience different fading conditions in the form of cooperative diversity. Another challenge is insufficient infrastructure in remote areas. Many highways and railways pass through under-populated areas where BSs and RSUs are rare and new ones are difficult or costly to build. Relaying has been demonstrated to increase energy efficiency and extend coverage in cellular systems. In the meantime, it has been shown that vehicles tend to form clusters, which makes it possible for them to participate as cooperative relays in communications. Hence, it is a straightforward idea to employ relaying to extend the coverage of vehicular communications. Since vehicles are mostly rational or selfish mobile users (different from the BSs or relay stations operated by operators), EH relaying technique actually provides them an incentive to participate in such cooperative relaying without consuming their own power.

Given the aforementioned benefits of EH relaying, it has attracted many research efforts in recent years. Among these work, SWIPT is a crucial technique that enables wireless receivers to both harvest energy and extract information from wireless

Table 3.2 Related work EH relay networks

	SWIPT method	Relaying mode	Communication directions	Relay selection	Relay protocol
[35]	TS & PS	HD	One-way	None	AF
[36]	TS	FD	One-way	None	AF & DF
[37]	N/A	FD	One-way	None	AF
[38]	PS	HD	One-way	SRS	AF
[39]	PS	FD	One-way	None	AF
[40]	PS	FD	Two-way	GRS	AF

signals. Unfortunately, a practical limitation of all SWIPT devices is that the EH relays are not yet able to directly process the information carried by the signals [33]. Consequently, coordination between EH and information processing (IP) is necessary by either a time-switching-based relaying (TSR) or power-splitting-based relaying (PSR) protocol.

An overview of existing work on EH relaying is presented in Table 3.2. Next, we will review these different schemes from two perspectives: (i) Relay operation mode in terms of half duplex (HD) or full duplex (FD); (ii) Relay participation mode in terms of random or optimally selected relay engagement.

3.5.2.1 HD and FD EH Relaying

Conventionally, most work in the literature focus on HD EH relays, where each transmission cycle is divided into two phases. The relay receives signals from the source, with SWIPT, in the first phase and forwards the signals to the destination in the second one, using energy harvested in the first phase. Recently, the throughput of HD relay networks with SWIPT at the relay was studied.

Since the breakthrough in hardware and/or software based self-interference mitigation methods, FD relay networks has been under the spotlight of researchers because of their significant performance advantages over HD ones in many scenarios. In these systems, the relays are capable of FD operation, i.e., receiving and transmitting signals simultaneously over shared radio resources. In theory, the throughput can be nearly doubled similar to conventional relay networks without EH. With practical issues considered, including the self-interference and different TS/PS designs, the performance of FD EH relay networks need to be investigated more carefully.

A TS-based FD relaying scheme, referred to as FD-TSR-I in this section, was proposed in [36]. In this scheme, a TS factor $\alpha \in [0, 1]$ is employed. Each transmission cycle of length T is divided into two phases according to α. Only EH is conducted at the relay in the first phase, $t \in [0, \alpha T)$. In the second phase, $t \in [\alpha T, T)$, the relay receives and processes information from the source and forwards to the destination simultaneously, with either amplified-and-forward (AF) or decode-and-forward (DF) relaying protocols. The globally optimal TS factor

α, which should maximize the network throughput, can be obtained by univariate optimization techniques.

In [37], another FD EH relaying scheme, referred to as FD-TSR-II in this section, was proposed with two evenly divided phases in each transmission cycle. In contrast to FD-TSR-I, only IP is conducted at the relay in the first half cycle. In the second half cycle, the energy carried by signals transmitted from the source is harvested while the relay forwards the information received in the first phase over the R-D link. Consequently, the loopback self-interference actually benefits the performance of the network, since the relay does not receive information in the second phase.

In [39], a PS-based FD EH relaying scheme, referred to as FD-PSR in this section, was proposed. In this scheme, the relay simultaneously transmits and receives information without interruption. At the relay, the received signals are split for EH and IP respectively according to a PS factor ρ.

The differences between the schemes mentioned above are depicted in Fig. 3.15. The relay in both FD-TSR-I and FD-TSR-II systems can only transmit and/or receive information during a fraction of time in each transmission cycle. The remaining part of the transmission cycle needs to be allocated for EH. This grants PS-based FD EH relaying schemes a unique advantage.

The end-to-end capacity of these three schemes are compared with via simulations with parameters set as follows. The source transmit power is 30 dBm and the average path loss of S-R and R-D channels is −60 dB. All channels responses are assumed to follow Rayleigh fading with a noise power of −90 dBm. Note that FD-TSR-II outperforms the other two schemes under stronger loop-

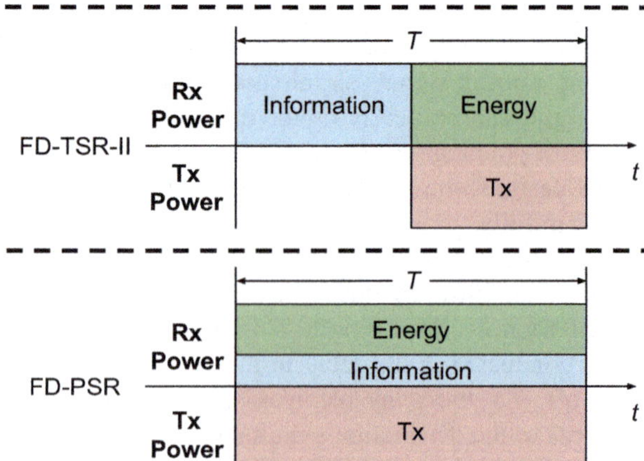

Fig. 3.15 FD EH relaying schemes

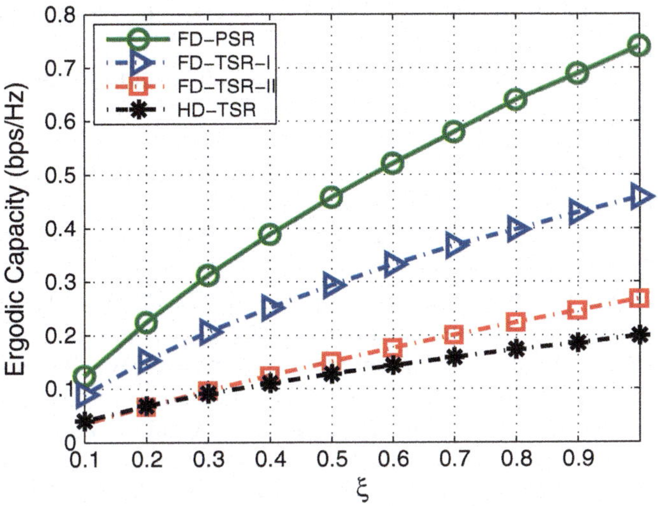

Fig. 3.16 Capacity comparisons among HD and FD EH relaying schemes

back self-interference, while FD-TSR-I and FD-PSR perform better under weaker loopback self-interference. Therefore, to ensure fair comparisons, the loopback self-interference channel gain is set to −15 dB for FD-TSR-II and −60 dB for FD-TSR-I and FD-PSR. The results are presented in Fig. 3.16, as well as their performance gains with respect to the corresponding HD networks. While all of them outperform the HD TS-based relaying, FD-PSR achieves a significantly higher average capacity than FD-TSR-I and FD-TSR-II.

3.5.2.2 EH Relay Selection

In practical wireless networks, one can employ multiple relays instead of installing multiple antennas on a device. This is particularly attractive for vehicular solutions, for which multiple antennas are currently impractical. To exploit the spatial diversity and further improve the system performance, relays should be judiciously selected, as opposed to randomly chosen. Hence, the relay selection (RS) problem is attracting considerable attention in the academia in recent years.

For conventional relay networks, most existing work on RS assumes that orthogonal channels are assigned to selected relays with independent power constraints. In this case, selecting a single relay clearly achieves the highest spectrum efficiency, since the relay with the strongest channel can occupy the entire bandwidth. In a shared bandwidth setup, however, single relay selection (SRS) is not always optimal, since multiple relays can enhance the SNR at the receiver via network beamforming. From another perspective, however, a larger number of selected relays leads to higher network energy consumption. When considering the EE of the network, the RS problem is further complicated.

The RS problems in various EH relay network architectures were proposed and investigated in the literature (Fig. 3.17). The ones with wireless powered relays (WPRs) have attracted most attention. In these network architectures, the WPRs are solely powered by EH from wireless signals. In other words, they do not consume their own power to process and forward signals.

In the networks with WPRs, the energy harvested from the received signals, instead of that stored in batteries, is used for relaying. In [38], The SRS problem in HD PS-based EH relay networks was investigated. The PS factor for each WPR is optimized and the RS objective is to minimize the outage probability of the network. The RS problem was extended by considering the possibility of selecting multiple WPRs in [40], in which the RS problem in FD EH relay networks was investigated. In contrast to conventional relay networks, selecting more WPRs helps harvest more energy that can be used for relaying instead of increasing the network energy consumption. For FD networks, however, bandwidth splitting is still necessary when multiple relays are selected, which may degrade the throughput. Therefore, neither SRS or all-participate RS is universally optimal.

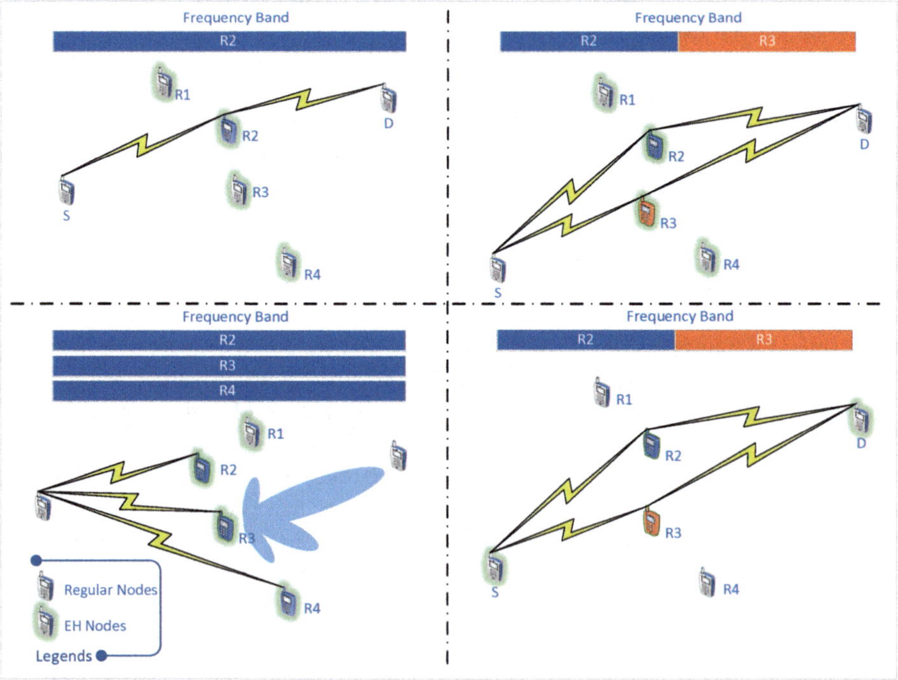

Fig. 3.17 Relay selection in various EH relay networks

On the other hand, the general RS (GRS) problem is NP-hard. A greedy algorithm was henceforth proposed in [40, 41] that yields near-optimum performance. The proposed greedy RS method starts with an empty relay set $\mathcal{R}_0 = \varnothing$ and gradually adds the relays one by one. At each step, we maintain the selected relays in the previous step and add one of the remaining relays that would achieve the maximum capacity if it were added to the selected relay set in the previous step. Let $\mathcal{R}_i$ denotes the selected relays in the ith step. Hence, $\mathcal{R}_0 \subset \mathcal{R}_1 \subset \cdots \subset \mathcal{R}_N$. The intuition that inspired the algorithm lies in that the relays selected in the previous step, with fewer relays cooperating, are expected to experience better channel conditions, to the source and the destination. Other combinations of the same number of relays are unlikely to outperform this set of relays when adding an extra relay. Therefore, the relays selected in the previous step should be selected with priority in the current step as well. Although this is a heuristic algorithm, i.e., the optimality is not guaranteed, simulation results given in Fig. 3.18 demonstrate that its performance is very close to the exhaustive-search-based optimal algorithm. In addition, as shown in Fig. 3.18, the all-participate (AP) method that involves all the available relays into the transmission and the SRS method that restricts the number of selected relays to one approach the optimal performance at the lower and higher ends of the SNR axis, respectively. This phenomenon can be explained in the following. Having more relays cooperating leads to more overall harvested energy for relaying. This is desirable when the source transmit power is low. On the other hand, selecting fewer relays results in larger bandwidth for each of them. This

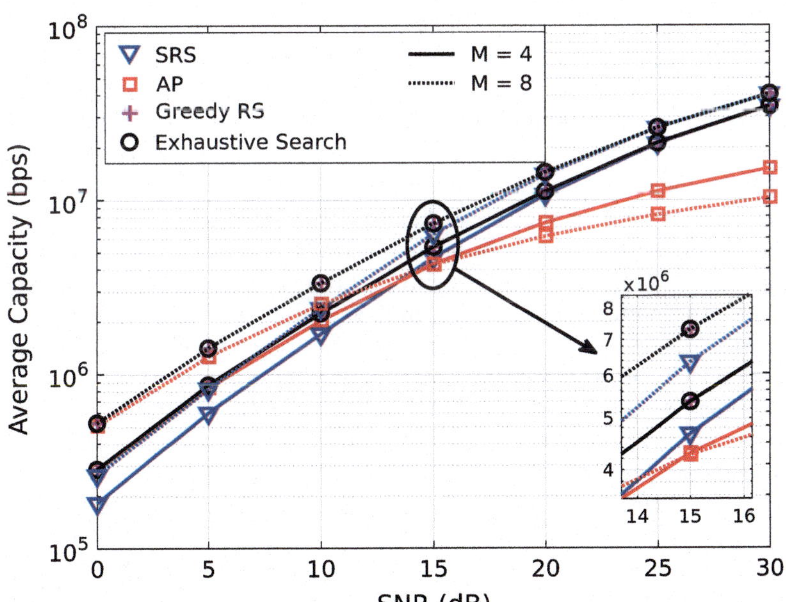

Fig. 3.18 Average capacity comparison of different EH RS methods

is a more efficient way to utilize the spectrum while the source transmit power is relatively high. This result also demonstrates that in FD EH relay networks, SRS cannot always lead to the optimal performance like in most conventional relay networks.

In the research work mentioned above, the signal processing and forwarding always entirely consumes the harvested energy in the current transmission cycle. With energy storage considered, the relays have the power management flexibility. The RS problem in such systems was studied in [42], where each relay is equipped with a battery and participates as an RS candidate only if its battery is fully charged. The significant performance gain was verified by simulations.

Besides RS in EH networks with WPRs, some other less common network architectures in which EH occurs at sources and/or destinations have also been considered. For example, RS in the networks with one information destination and multiple energy destinations was investigated. Another interesting network configuration is to employ conventional non-EH relays to facilitate one-way or two-way communications between nodes, one or both of which are energy constrained and can harvest energy [43]. However, the RS problem in such networks remains uncharted. Compared with random relay deployment, selecting the optimal relay could utilize the spectrum most efficiently. Performance can also be improved by exploiting the spatial diversity.

3.6 The Next Leap

In the current state of the art, PHY techniques for vehicular communications are mainly designed to combat the severe channel conditions. For instance, when OFDM is employed, efficient ICI cancellation methods are then required to alleviate the harmful effect of the fast time-varying channel. As the next leap, one would head to shift from combating channel to exploiting channel.

The vehicular communications channels that inherently comes with highly dynamic Doppler effects are even non-stationary. Therefore, the widely used OFDM technique may not be the optimal multi-carrier transmission scheme in these fast time-varying scenarios. It is of great significance to find a more robust multi-carrier transmission scheme, which possesses good time-frequency (T-F) localization. Furthermore, the transmission scheme should be flexibly adjusted based on the channel characteristics. Noting that the double selectivity also brings double diversity in the T-F domain, the flexible index modulation within the two-dimensional T-F resource grid is also another promising technique that is worth investigation.

Distributed MIMO design via EH relays in vehicular networks can provide positive spatial diversity and improve the spectral efficiency for V2X communications. The efficient and effective cooperation mechanism should be investigated for the optimal design of distributed MIMO in vehicular networks. Moreover, massive MIMO has been regarded as one of the core techniques in 5G, which can significantly increase the multi-user data rates. However, in vehicular communications, due to the imperfect CSI, the application of massive MIMO is facing severe challenges. How to alleviate the effect of imperfect CSI while exploiting the benefits of massive MIMO in vehicular communications still remains an open problem.

References

1. Wireless LAN Medium Access Control (MAC) and Physical Layer (PHY) Specifications Amendment 6: Wireless Access in Vehicular Environments, IEEE Standard 802.11p, 2010.
2. S. Chen, J. Hu, Y. Shi, and L. Zhao, "LTE-V: A TD-LTE-Based V2X Solution for Future Vehicular Network," *IEEE Internet of Things Journal*, vol. 3, no. 6, pp. 997–1005, Dec. 2016.
3. S. Chen *et al.*, "Vehicle-to-Everything (V2X) Services Supported by LTE-Based Systems and 5G," *IEEE Communications Standards Magazine*, vol. 1, no. 2, pp. 70–76, 2017.
4. M. Wen, X. Cheng, X. Wei, B. Ai, and B. Jiao, "A novel effective ICI self-cancellation method," in *Proc. IEEE GLOBECOM 2011*, Dec. 2011.
5. M. Wen, X. Cheng, L. Yang, and B. Jiao, "Two-path transmission framework for ICI reduction in OFDM systems," in *IEEE Global Communications Conference (GLOBECOM 2013)*, Atlanta, GA, 2013, pp. 3716–3721.
6. X. Cheng, Q. Yao, M. Wen, C. X. Wang, L. Song, and B. Jiao, "Wideband channel modeling and intercarrier interference cancellation for vehicle-to-vehicle communication systems," *IEEE Journal on Selected Areas in Communications*, vol. 31, no. 9, pp. 434–448, Sept. 2013.
7. H. G. Yeh and C. C. Wang, "New parallel algorithm for mitigating the frequency offset of OFDM systems," in *Proc. IEEE VTC'04-Fall*, Los Angeles, USA, Sep. 2004, pp. 2087–2091.
8. H. G. Yeh, Y. K. Chang, and B. Hassibi, "A scheme for cancelling intercarrier interference using conjugate transmission in multicarrier communication systems," *IEEE Trans. Wireless Commun.*, vol. 6, no. 1, pp. 3–7, Jan. 2007.
9. Y. Zhao, J. D. Leclercq, and S. Häggman, "Intercarrier interference compression in OFDM communication systems by using correlative coding," *IEEE Commun. Lett.*, vol. 2, no. 8, pp. 214–216, Aug. 1998.
10. Y. Zhao and S. G. Häggman, "Intercarrier interference self-cancellation scheme for OFDM mobile communication systems," *IEEE Trans. Commun.*, vol. 49, no. 7, pp. 1185–1191, Jul. 2001.
11. Y. Fu, S. G. Kang, and C. C. Ko, "A New Scheme for PAPR Reduction in OFDM Systems with ICI Self-Cancellation," in *Proc. IEEE VTC'02-Fall*, Vancouver, Canada, Sep. 2002, pp. 1418–1421.
12. K. Sathananthan, R. M. A. P. Rajatheva, and S. B. Slimane, "Cancellation technique to reduce intercarrier interference in OFDM," *Electron. Lett.*, vol. 36, no. 25, pp. 2078–2079, Dec. 2000.
13. K. Sathananthan, C. R. N. Athandage, and B. Qin, "A Novel ICI Cancellation Scheme to Reduce both Frequency Offset and IQ Imbalance Effects in OFDM," in *Proc. IEEE ISCC'04*, Alexandria, Egypt, Jul. 2004, pp. 708–713.
14. C. L. Wang and Y. C. Huang, "Intercarrier interference cancelling using general phase rotated conjugate transmission for OFDM systems," *IEEE Trans. Commun.*, vol. 58, no. 3, pp. 812–819, Mar. 2010.
15. E. Basar, U. Aygolu, E. Panayirci, and H. V. Poor, "Orthogonal frequency division multiplexing with index modulation," *IEEE Trans. Signal Process.*, vol. 61, no. 22, pp. 5536–5549, Nov. 2013.
16. X. Cheng, M. Zhang, M. Wen, and L. Yang, "Index Modulation for 5G: Striving to Do More with Less," *IEEE Wireless Commun. Mag.*, 2017.
17. Y. Li, M. Zhang, X. Cheng, M. Wen and L. Yang, "Index modulated OFDM with intercarrier interference cancellation," in *Proc. 2016 IEEE Int. Conf. on Commun. (ICC)*, Kuala Lumpur, 2016, pp. 1–6.

18. Y. Li, M. Wen, X. Cheng and L. Yang, "Index Modulated OFDM with ICI Self-Cancellation," in *Proc. 2016 IEEE 83rd Veh. Technol. Conf. (VTC Spring)*, Nanjing, 2016, pp. 1–5.
19. Y. Li, M. Wen, X. Cheng and L. Yang, "Index modulated OFDM with ICI self-cancellation for V2X communications," in *Proc. 2016 Int. Conf. Comput. Networking and Commun. (ICNC)*, Kauai, HI, 2016, pp. 1–5.
20. G. Acosta-Marum and M. A. Ingram, "Six time- and frequency-selective empirical channel models for vehicular wireless LANs," *IEEE Veh. Tech. Mag.*, vol. 2, no. 4, pp. 4–11, Dec. 2007.
21. Y. Bian, X. Cheng, M. Wen, L. Yang, H. V. Poor and B. Jiao, "Differential Spatial Modulation," *IEEE Transactions on Vehicular Technology*, vol. 64, no. 7, pp. 3262–3268, July 2015.
22. M. Zhang, X. Cheng, and L. Yang, "Differential Spatial Modulation in V2X," in *Proceedings of the 9th EUCap*, Lisbon, Portugal, Apr. 2015.
23. M. Renzo, H. Haas, and P. Grant, "Spatial modulation for multiple-antenna wireless systems: A survey," *IEEE Commun. Mag.*, vol. 49, no. 12, pp. 182–191, Dec. 2011.
24. J. Proakis, *Digital Communications*, 4th Edition. New York: McGraw-hill, 2001.
25. V. Tarokh and H. Jafarkhani, "A differential detection scheme for transmit diversity," *IEEE J. Select. Areas Commun.*, vol. 18, no. 7, pp. 1169–1174, Jul. 2000.
26. http://en.wikipedia.org/wiki/Stirling's_approximation
27. N. Serafimovski, M. Renzo, S. Sinanovic, R. Mesleh, and H. Haas, "Fractional bit encoded spatial modulation (FBE-SM)," *IEEE Commun. Lett.*, vol. 14, no. 5, pp. 429–431, May 2010.
28. X. Zhou, R. Zhang, and C. K. Ho, "Wireless information and power transfer: Architecture design and rate-energy tradeoff," *IEEE Trans. Commun.*, vol. 61, no. 11, pp. 4754–4767, Nov. 2013.
29. L. R. Varshney, "Transporting information and energy simultaneously," in *Proc. 2008 IEEE International Symposium on Information Theory*, Toronto, ON, 2008, pp. 1612–1616.
30. P. Grover and A. Sahai, "Shannon meets Tesla: Wireless information and power transfer," in *Proc. 2010 IEEE International Symposium on Information Theory*, Austin, TX, 2010, pp. 2363–2367.
31. R. Zhang and C. K. Ho, "MIMO Broadcasting for Simultaneous Wireless Information and Power Transfer," *IEEE Transactions on Wireless Communications*, vol. 12, no. 5, pp. 1989–2001, May 2013.
32. R. Atallah, M. Khabbaz and C. Assi, "Energy harvesting in vehicular networks: A contemporary survey," *IEEE Wireless Communications*, vol. 23, no. 2, pp. 70–77, April 2016.
33. D. Wang, R. Zhang, X. Cheng, Z. Quan, and L. Yang, "Joint Power Allocation and Splitting (JoPAS) for SWIPT in Doubly Selective Vehicular Channels," *IEEE Trans. Green Commun. Netw.*, vol. 1, no. 4, pp. 494–502, Dec. 2017.
34. S. Boyd and L. Vandenberghe, Convex optimization. Cambridge University Press, 2004.
35. A. A. Nasir, X. Zhou, S. Durrani and R. A. Kennedy, "Relaying Protocols for Wireless Energy Harvesting and Information Processing," *IEEE Transactions on Wireless Communications*, vol. 12, no. 7, pp. 3622–3636, July 2013.
36. C. Zhong, H. A. Suraweera, G. Zheng, I. Krikidis and Z. Zhang, "Wireless Information and Power Transfer With Full Duplex Relaying," *IEEE Transactions on Communications*, vol. 62, no. 10, pp. 3447–3461, Oct. 2014.
37. Y. Zeng and R. Zhang, "Full-Duplex Wireless-Powered Relay With Self-Energy Recycling," *IEEE Wireless Commun. Lett.*, vol. 4, no. 2, pp. 201–204, 2015.
38. K. Liu, "Outage-optimal relay selection for energy-harvesting relays based on power splitting," in *Proc. 2015 International Conference on Wireless Communications & Signal Processing (WCSP)*, Nanjing, 2015, pp. 1–6.

39. D. Wang, R. Zhang, X. Cheng, and L. Yang, "Capacity-Enhancing Full-Duplex Relay Networks based on Power-Splitting (PS-)SWIPT," *IEEE Trans. Veh. Technol.*, vol. 66, no. 6, pp. 5445–5450, Jun. 2017.
40. D. Wang, R. Zhang, X. Cheng, L. Yang, and C. Chen, "Relay Selection in Full-Duplex Energy-Harvesting Two-Way Relay Networks," *IEEE Trans. Green Commun. Netw.*, vol. 1, no. 2, pp. 182–191, Jun. 2017.
41. D. Wang, R. Zhang, X. Cheng, and L. Yang, "Full-Duplex Energy-Harvesting Relay Networks: Capacity-Maximizing Relay Selection," *Journal of Communications and Information Networks*, 2018, to appear.
42. I. Krikidis, "Relay Selection in Wireless Powered Cooperative Networks With Energy Storage," *IEEE Journal on Selected Areas in Communications*, vol. 33, no. 12, pp. 2596–2610, Dec. 2015.
43. J. Rostampoor, S. M. Razavizadeh, and I. Lee, "Energy Efficient Precoding Design for SWIPT in MIMO Two-Way Relay Networks," *IEEE Transactions on Vehicular Technology*, vol. 66, no. 9, pp. 7888–7896, Sept. 2017.
44. http://en.wikipedia.org/wiki/Factorial_number_system

Chapter 4
Wireless-Vehicle Combination: Effective MAC Designs in VCN

In vehicular networks, road safety and data-related applications require reliable and efficient communications with minimized transmission collisions, and thus effective medium access control (MAC) protocols are also essential for the VCN system design. However, in vehicular networks, the MAC design is much more challenging due to the high mobility, heterogeneous and frequently changing topology, and versatile QoS requirements. Hence, also following the wireless-vehicle combination perspective, in this chapter, we focus on the effective and efficient MAC designs for VCN. First, by taking the trend (from distributed to centralized) and challenges of the MAC design in VCN into consideration, three essential MAC issues and their corresponding solutions are explored, including distributed congestion control, centralized resource sharing and scheduling, and centralized data dissemination scheduling. Then, regarding the next leap for MAC designs in VCN, NOMA-V2X and data-driven resource management are also discussed.

4.1 MAC Designs in VCN

In the literature, the MAC design in vehicular networks can be categorized into two classes, namely distributed MAC protocols and centralized MAC protocols.

The enhanced distributed channel access (EDCA) protocol based on IEEE 802.11p [1] and 802.11e [2] is the most widely applied distributed MAC design in vehicular networks. The EDCA protocol differentiates packets from an upper layer into four different access categories (ACs) based on the quality-of-service (QoS) requirement of the applications at higher layers. The four available traffic categories corresponding to the ACs are: VOice traffic (VO), VIdeo traffic (VI), Best Effort traffic (BE), and BacKground traffic (BK). Each AC works as an independent vehicle with the enhanced distributed channel access function (EDCAF) to contend for the transmission opportunities (TXOP) using its own EDCA parameters, that is, the arbitration inter-frame space (AIFS), the minimum contention window size

X. Cheng et al., *5G-Enabled Vehicular Communications and Networking*, Wireless Networks, https://doi.org/10.1007/978-3-030-02176-4_4

(CW_{min}), and the maximum contention window size (CW_{max}). In the EDCA protocol, if the channel is sensed idle for the duration of AIFS[AC] and there is backlogged data in the AC, the backoff counter for the EDCAF will be started and randomly chooses a number in the range $[0, CW[AC]]$ as a starting point of countdown, where $CW[AC]$ denotes the contention window size of the AC and is initially set as $CW_{min}[AC]$. If the medium becomes busy in the backoff process, the backoff counter will be frozen and resumed until the channel is sensed idle again for more than a AIFS. When the backoff counter reaches zero, the AC will initiate a transmission sequence. If the transmission attempt of the AC fails, $CW[AC]$ will be doubled till the corresponding $CW_{max}[AC]$. Once the data of the AC transmits successfully, $CW[AC]$ will be set back to $CW_{min}[AC]$ again. The EDCA protocol can be easily applied in vehicular ad-hoc networks (VANETs) and achieves efficient distributed transmission scheduling.

Although distributed MAC protocols are readily applicable in vehicular networks with a distributed topology, they usually have limited network throughput and would experience severe data congestion especially when the traffic is heavy. On the other hand, centralized MAC protocols make more optimized scheduling decisions based on the collected information, leading to reduced data transmission conflicts and improved network throughput. In 5G-based C-V2X networks, low-latency and high-rate V2X communications make the CSI and vehicle information collection at the central entities as well as the control information exchange more efficient, and thus will also increase the feasibility and efficiency of centralized MAC protocols. Therefore, in terms of the MAC design in vehicular networks, there is an evident trend to go from the easily applicable distributed protocols to the more effective centralized ones.

In the following, three essential MAC issues and their corresponding solutions will be discussed, including distributed congestion control, centralized resource sharing and scheduling, and centralized data dissemination scheduling. The overall structure of this section is given in Fig. 4.1.

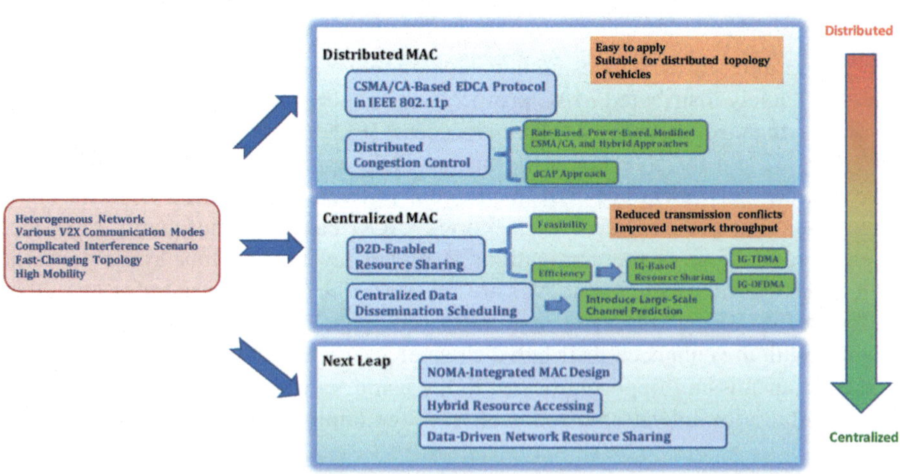

Fig. 4.1 Content structure of 5G-VCN MAC design

4.2 Distributed Congestion Control

According to the EDCA protocol, there are four EDCAFs inside a vehicle, and thus it is possible that more than one ACs initiate a transmission sequence at the same time. Therefore, a collision may occur inside a single vehicle. A scheduler inside vehicles will avoid this kind of internal collisions by granting the TXOP to the highest priority traffic. However, When there are more than one ACs are granted TXOP by different vehicles, an external collision occurs and cannot be avoided since there is no priority among vehicles. Especially when there are a large number of vehicles requesting to send messages at the same time, the shared radio medium could be easily congested, leading to significantly compromised system throughput and reliability. Therefore, congestion control is a critical problem in vehicular networks, especially for current IEEE 802.11p-based vehicular communications which lacks the centralized coordination of the data transmission requests. The congestion control approach is then put forward to avoid the high external collision problem in a dense vehicular network.

In this section, we first provide an overview of the existing congestion control approaches in vehicular networks and then emphasize on introducing a distributed congestion-adaptive priority (dCAP) approach that achieves improved congestion control performance.

4.2.1 Overview of Existing Congestion Control Approaches

In the literature, many work have contributed to the congestion control for the IEEE 802.11p-based vehicular communications. The objective of congestion control is to limit the high load on the shared communication channel and provide efficient and fair channel access for vehicles. Therefore, congestion control approaches typically dynamically adjust one or more transmission parameters, such as transmission rate, packet generating rate, transmit power, transmission range, and the parameters in the EDCA protocol, according to the measured channel congestion condition. Generally, existing congestion control approaches can be classified into four categories, that is, rate adaption-based, power adaption-based, modified CSMA/CA protocol-based, and hybrid approaches [3].

1. Rate adaption-based approach: Adjust the transmission rate or packet generating rate according to the network density.
2. Power adaption-based approach: Adjust the transmission power or transmission range in response to the channel load.
3. Modified CSMA/CA protocol-based approach: Modify the control parameters of the CSMA/CA protocol to alleviate channel collision, for instance increasing the CW size properly to alleviate channel collision.
4. Hybrid approach: Jointly consider the rate adaption, power adaption, and modified CSMA/CA protocol based approaches.

4.2.1.1 Rate Adaption-Based Approach

The rate adaption-based congestion control approaches [5–9] aim to reduce the channel collisions through adjusting the transmission rate or packet generating rate according to the network density. The rate adaption methods are usually implemented upper the MAC layer and based on the transmission statistics collected by the lower layers, i.e., MAC and physical layers.

In [5], the utility-based packet forwarding and congestion control (UBPFCC) algorithm was proposed. The packet priority is calculated as a function of packet utility and packet size. For a utility-fair rate assignment, the transmission rate is proportional to the priority and effective data rate. For the collision avoidance purpose, the transmission rate is disproportional to the estimated number of nodes in the transmission range. In case of congestion, packets with a low utility for the network are dropped. The UBPFCC algorithm can effectively protect the packet transmission with high utility and provide fair transmission opportunities among packets with equal utility. However, the information of priority and effective data rate exchange results in high system overhead.

In [6], the authors proposed an adaptive rate control algorithm based on network condition and tracking error for safety applications in vehicular ad-hoc networks. The proposed on-demand rate control (ODRC) algorithm calculates transmission probability based on suspected tracking error on neighboring vehicles toward its own position in Euclidean sense. The transmission probability increases with the suspected tracking error due to the fact that a higher transmission probability is required for a vehicle that has more unexpected movements. On the other hand, the transmission probability decreases with the time-averaged channel collision intensity. Since there is no explicit receive acknowledgement (ACK) for the broadcasting transmission, the suspected tracking error is decided based on the packet erasure rate and the channel collision intensity is substituted by the local channel utilization. The ODRC method works in a fully distributed manner without any interactions among vehicles and effectively improves the tracking performance. However, unlike [5], the packet priority is out of consideration here.

In [7], it studied the trade-off between the broadcasting efficiency and reliability and obtained the optimal packet transmission probability corresponding to the vehicle density. Then three congestion control approaches were proposed to achieve the maximum broadcasting efficiency. The first one is a direct MAC layer congestion control approach based on the contention window size adaption corresponding to the optimal packet transmission probability. The second one is a cross layer approach with the MAC signaling for sending a packet to the short message layer with a probability to achieve the optimal packet transmission probability. The third one is a simple rate control outputting the desired transmission data rate according to the vehicle density. Work [7] makes outstanding contributions on the selection of optimal packet transmission probability to maximize the broadcasting efficiency. However, the investigated communication scenario, where all vehicles are randomly positioned in a single lane modeled as one-dimensional (1-D) spatial network, is not a general case.

In [8], the authors proposed a probabilistic discard congestion control for safety information in V2I vehicular network. In the proposed congestion control mechanism, the non-safety related packets are accepted with a defined probability when the size of queue within the RSU is larger than a threshold. The approach can make the packet dropping probability of the safety-related packets lower than that of the non-safety related packets. However, the article takes no consideration of the basic system channel condition based on which the threshold may be adjusted.

In [9], another traffic rate adaption method was introduced for the congestion control. In the proposed method, the transmission rate of the periodic safety application (PSA) is initially set to R_0 and increases with a fast time step. This behavior continues until the rate reaches the rate limit or a congestion event occurs. A congestion event is supposed to happen when the MAC blocking event happens. When a congestion event occurs, half of the current traffic rate is saved as the fast start threshold and fast start begins again from R_0 after PSA messages are unblocked at the MAC layer. Once the traffic rate reaches the threshold, the rate control goes into the congestion avoidance mode, where the traffic rate increases with a small time step. The goal of the PSA rate adaption is to protect the transmission for the event driven safety application (ESA) with higher priority than the PSA. However, in addition to just decreasing the rate threshold, the proposed method can be further improved by increasing the threshold when the channel load is released.

4.2.1.2 Modified CSMA/CA Protocol-Based Approach

The modified CSMA/CA protocol-based congestion control approaches [10–15] adopt a direct manner by manipulating the channel access control parameters response to the channel congestion condition. The parameter most frequently modified for the congestion control is the contention window size.

In [10], the authors proposed a novel approach called adaptive offset slot (AOS) to decrease the channel load based on the idea that there is an optimal minimum contention window size corresponding to the number of neighbor vehicles. In the AOS approach, the minimum contention window size is plus by an offset slot according the result of contestant estimation. The performances of throughput and packet loss probability get significant improvements with the AOS method. However, the article takes no further consideration of the condition when the congestion is released as well as [9].

In [11], it provided a detection-based MAC mechanism by modifying RTS/CTS to detect network congestion through message exchange and predict the number of competing nodes. The contention window size is adjusted according to the number of competing nodes to maximize the capacity performance. Specifically, the contention window size increases with the number of competing nodes to avoid the channel overload problem. However, this proposed method just works on the condition that all packets are of the same service level.

In [12], the authors proposed two concrete congestion control approaches to guarantee the transmission for vehicular safety applications. The first one is that once a safety packet is generated, all the non-safety transmissions should be frozen except the safety transmissions. The other one is that the contention window size of the non-safety transmissions is dynamically adjusted according to the periodically measured channel usage. If the measured channel usage is over a queue freeze threshold, the non-safety transmissions will also be frozen. Otherwise, if the channel usage is just over a congestion threshold, which is smaller than the queue freeze threshold, the contention window size is doubled till the maximum value. On the other hand, if the channel usage is lower than the sparse threshold, the contention window size is then divided by 2 to improve the channel utilization. The article aims to protect the transmission of the safety applications while sacrificing the performance of the non-safety applications. However, The proposal in [12] ignores the collision problem among safety applications, which is the main reason leading to the performance degradation of the high priority traffic flaws in the IEEE 802.11p vehicular network.

In addition to the contention window size, the parameter AIFS could also be used for the collision avoidance as in [13]. The authors in [13] established strict priorities in IEEE 802.11p vehicular networks by enlarging the AIFS of a given access category with the largest possible backoff interval of all access categories corresponding to higher-priority frames. The proposed method reduces the collision rate between the higher priority traffics at the expense of increasing the delay, especially for the lower priority traffics.

Other than adjusting the channel access parameters of MAC layer, the authors in [14, 15] modified the channel access manner of MAC layer for the congestion control. In [14], the congestion control approach grants the channel access chance to the node with the highest transmission priority in its interference range. The transmission priorities of each node are exchanged through the beacon structure. A node can use the available bandwidth only if it holds the highest priority message, otherwise it freezes its sending. The proposed method in [14] ensures the channel access chance for the highest priority transmissions at the expense of system overhead. The authors in [15] proposed a Safety Range CSMA (SR-CSMA) protocol which aims to increase the reception probability for the safety messages in the immediate neighborhood and reduce the update delay between closely situated nodes. This proposal is based on the modification of physical carrier sensing mechanism, which allows simultaneous transmissions from distant nodes if satisfying the target Signal to Interference Ratio (SIR) thresholds requirement in the safety range. This approach enables the channel with the capability of accommodating an increased number of nodes. Therefore, this method keeps a high beaconing delivery ratio in the immediate neighborhood and preserves the efficiency of the safety applications. However, the proposal in [15] just considers simple case with one dimension network. Since the real network is more complex resulting in high difficulty for the SIR estimation, it may need more overhead for the transmission negotiation.

4.2.1.3 Power Adaption-Based Approach

The power adaption based congestion control approaches [16–19] dynamically adjust the transmission power or the transmission range to avoid the problem of channel overload. The amount of competing nodes in the same communication range automatically decreases when the transmission power is reduced and thus the channel collisions can be avoided. However, reducing transmission range can cause network disconnection and increases the number of hops which can have a negative effect on the end to end delay performance.

A fair power control for safety-critical information based on a strict fairness criterion, i.e., distributed fair power adjustment for vehicular environments (D-FPAV) was proposed in [16]. The proposed D-FPAV algorithm is actually based on a centralized FPAV algorithm [26] that node power levels are iteratively increased by the same amount and start from the minimum level. This process is continued as long as the condition on the maximum beacon load is satisfied. The D-FPAV algorithm is then designed with the following factors: (1) executing the FPAV algorithm at each node with the information gathered from received beacons; (2) exchanging the locally computed transmit power control values among surrounding vehicles; and (3) selecting the minimum power level among the one locally computed and those computed by the surrounding vehicles. The main goals of the D-FPAV algorithm are maximizing the minimum transmit power value assigned to the nodes with periodic beacon information and reserving the channel load for the event-driven emergency messages with higher transmission priority. The proposed algorithm achieves good fairness for the beacon information transmission. However, the exchange of transmit power information among nodes results in high system overhead.

In [17], the authors presented a congestion method based on the channel occupancy measurements. It aims to achieve the optimal system performance represented by information dissemination rate, which turns out to be related with an optimal channel occupancy value, i.e., 0.7. A mathematical model is designed to present the channel occupancy as a function of the network density, transmission rate, and transmission range. Since the network density and transmission rate are usually unavailable, the article designs a transmission range adaption method to achieve the optimal channel occupancy value with high convergence rate and the ability of fast react to the change of network density and transmission rate. The proposed method is inherently robust and does not rely on the knowledge of the propagation model, road density, or rate of transmission of other vehicles. However, the transmission range adaption method should be modified if the mathematical model is designed with different transmission rates among each vehicles.

In [18], the authors provided a forward broadcast scheme for fast data dissemination scheme and a backward rebroadcast scheme for reliable 1-hop data dissemination. In the forward broadcast scheme, the node whose position is farthest from the sender and the received power which is over a decoding threshold is selected as the next forward. In the backward broadcast scheme, the transmission range is linearly adjusted according to the channel occupancy to avoid the channel congestion problem as well as [20]. The proposed scheme ensures the high delivery ratio and fast emergency messages dissemination while avoiding congestion at the expense of high system overhead for the distance detection.

In [19], the authors proposed a multi-metric unicast data dissemination scheme (MUDDS) based on the communication density (CD) metric and a dissemination metric named LA. The CD is defined as the product of the transmission range, message frequency, and the vehicle density. In the MUDDS, each node locally measures network Packet Reception Rate (PRR) which is inversely proportional to the CD. When the measured PRR is over the target PRR threshold, reduce the transmission range. Otherwise raise the transmission range. The LA is defined as the product of the distance and the link availability rate. To mitigate the negative effect caused by the transmission range reduction, the neighbor with the largest LA value is selected to forward the data. It seems that the MUDDS works out the collision and network disconnection problems simultaneously. However, the exact adaption value of the transmission range based on the measured PRR is not presented in the MUDDS.

4.2.1.4 Hybrid Approach

The hybrid congestion control approaches [20–23] focus on more than one factor that affect the channel congestion problem. The transmission rate, channel access control parameters, and the transmission power can be jointly controlled to avoid the channel collisions in the hybrid congestion control approaches.

Based on the aforementioned work [6], the authors proposed a joint rate-power control algorithm for the vehicular tracking problem caused by the congested random access channel in [20]. The transmission probability is controlled by the simplified rate control method based on [6]. Here the channel utilization is not considered in the rate control part. While the channel condition is further considered in the power control part. The transmission power is dynamically adjusted based on the measured average channel occupancy. If the channel occupancy is over the maximum threshold, the minimum transmission power level is assigned. While if the channel occupancy is smaller than the minimum threshold, the maximum transmission power level is assigned. Otherwise, the transmission power is linearly changed with the measured channel occupancy. The joint rate-power design is robust and can considerably reduces the tracking error. However, as well as [6], it still ignores the traffic priority differentiation.

In [21], a centralized adaptive congestion control method was proposed for the vehicular networks in road intersections. Unlike the traditional congestion control methods which are based on the mathematical model, the proposed method utilizes the offline simulator to determine the optimal configurations of the PSA traffic rate and MAC backoff exponent. The determination of the optimal configurations is to be done by the access point (AP) only according to the estimated vehicle number. The vehicles are then informed with the optimal configurations by the AP. It demonstrates that the system performance gets significantly improved with the optimal configurations. However, the article does not describe the offline simulator design in detail, e.g., whether the priority differentiation is considered in the simulator or not. Therefore, the accuracy of the offline simulator is doubtful.

A contextual communications congestion control approach for cooperative vehicular network [24] was proposed in [22]. The authors considered the cooperative awareness messages (CAM) broadcasting for the specific lane change assistance application. It aims that each vehicle involved in the lane change application can receive at least one CAM before its warning distance during a given time window, while satisfying the application reliability requirement p_{app}. The proposed method is based on the methodology in [25] that calculating the minimum transmission power required for successfully exchanging at least one CAM with probability p_{app} at a given distance and CAM transmission frequency. The authors in [22] further obtained the CAM communication configurations including the joint transmission power and rate settings that satisfy the application requirements. To avoid the channel congestion problem, each vehicle can utilize the specific positions of neighboring vehicles to reconfigure its application requirements and the resulting transmission parameters. The channel busy time is efficiently reduced with the contextual cooperative transmission approach.

In [23], a joint transmission power and beacon rate control approach was proposed to achieve the fairness among beacon messages and reserve the channel resource for the emergency messages as well as [16]. The proposed scheme for safety messages dissemination consists of three phases as follows: (1) priority assignment based on a static metric according to the content type and a dynamic metric according to the transmission emergency condition; (2) collision detection depending on the metrics including the average waiting time, collision rate, and beacon reception rate; (3) transmission power control updated based on the collision rate and changing condition of 1-hop neighbors, and beacon rate adjustment according to the bandwidth fare share. The congestion condition is determined when the distance between the congestion index vector and the average congestion condition is larger than a threshold. The proposed method considerably improves the emergency messages reception ratio at the expense of the time and overhead for the collision detection.

4.2.2 Distributed Congestion-Adaptive Priority (dCAP) Approach

Different from most existing congestion control approaches, in which the variety of traffic applications is not taken into consideration and frequent information exchange among vehicles is required, in [3, 4], a distributed congestion-adaptive priority (dCAP) approach was proposed, which can prioritize the ACs based on the local transmission statics without information exchange among vehicles. The dCAP approach is fully compatible with the IEEE 802.11p MAC design. Moreover, [3] pointed out that the collision problem among the high priority traffics is the main factor that deteriorates the transmission performance of the high priority traffic in dense vehicular networks. This, however, has long been overtook in most existing congestion control approaches.

4.2.2.1 Congestion Condition Measurement

When the radio channel gets congested, it will lead to the following two consequences: (1) the queue length of some ACs which cannot access the channel becoming longer, and (2) the transmission failure probability becoming larger due to the high collision probability. Therefore, an AC v with traffic priority level m can measure the congestion condition in the following manner

$$C_m(v) = \frac{N_{queue}(v) + N_{fail}(v)}{N_{total}(v)} \tag{4.1}$$

where $N_{fail}(v)$ is the number of packets transmitted unsuccessfully, $N_{total}(v)$ is the total number of packets the AC v have generated and $N_{total}(v) = N_{queue}(v) + N_{fail}(v) + N_{suc}(v)$, where $N_{suc}(v)$ is the number of packets transmitted successfully. According to (4.1), if $C_m(v)$ becomes larger, the congestion condition of the radio channel become more serious.

4.2.2.2 dCAP Algorithm

In the dCAP algorithm proposed in [3, 4], given M traffic priority levels, the AC v with priority level $m = 1, 2, \cdots, M$ in a vehicle periodically measures the congestion condition with (4.1), where the time interval of measurement is predefined as T. During a time interval T, AC v keeps counting $N_{total}(v)$, $N_{queue}(v)$, and $N_{fail}(v)$, and updates $C_m(v)$ at the end of each T. The congestion condition $C_m(v)$ is then compared with the threshold assigned for its corresponding priority level, and denoted as $C_{th}(m)$. If $C_m(v)$ is greater than $C_{th}(m)$, which means the radio channel gets congested, then the AC should increase CW by multiplying a scaling factor $a(=2)$. On the other hand, if C_m is less than $C_{th}(m)$, which means the channel congestion condition gets eased, then the AC should reduce CW by dividing a scaling factor $a(=2)$ till the initial value. In addition, at the end of each T, the three parameters $N_{total}(v)$, $N_{queue}(v)$, and $N_{fail}(v)$ are reset as zero for the channel congestion measurement of the next time interval T. Note that under the congested channel condition, the low priority traffic should reduce the transmission rate to give access chances up to the traffic with higher priority than theirs. Therefore, the lower the traffic priority m, the smaller value of the congestion threshold $C_{th}(m)$. The detailed procedure of the dCAP approach is illustrated in Fig. 4.2.

Obviously, the congestion threshold is close related to the vehicular network density. When the network density is low, the low threshold would lead to the

---

Distributed multi-priority congestion control approach at AC queue v with traffic priority m

---

Initialize the parameter $CW_{min}(m) = CW_{min}(m)_{init}$
if end of the time interval T **then**
  Update the parameter C_m indicating the congestion condition
  $C_m(v) = [N_{queue}(v) + N_{fail}(v)] / N_{total}(v)$
  if $C_m > C_{th}(m)$ **then**
    $CW_{min}(m) = min\{CW_{min}(m) \times 2, CW_{max}(m)\}$
  else
    $CW_{min}(m) = max\{CW_{min}(m)/2, CW_{min}(m)_{init}\}$
  end if
else
Use the previous $CW_{min}(m)$, keep observing
$N_{queue}(v) = N_{new\_queue}(v)$
$N_{fail}(v) = N_{fail}(v) + N_{new\_fail}(v)$
$N_{total}(v) = N_{total}(v) + N_{new\_genert}(v)$
end if

Fig. 4.2 The detailed procedure of the dCAP approach

low system throughput since each AC may increase the CW unnecessarily. On the other hand, when the network density is high, the high threshold would lead to the frequent collisions since each AC does not increase the CW properly. Therefore, the congestion threshold should be dynamically adjusted with the vehicular network density, i.e., $C_{th}(m) = f(density)$.

4.2.2.3 Performance Evaluation

The optimized congestion thresholds over the vehicular network density are demonstrated via simulations in Fig. 4.3. The network density is represented by the number of vehicles competing the same channel resource. The optimized congestion threshold is determined with the principle that guaranteeing the transmission chance for the highest priority traffic as much as possible while restricting the collision probability under a determined constraint value in the simulation, i.e., 5%. From Fig. 4.3, it can be seen that the congestion threshold of each AC decreases with the number of vehicles. When the vehicular network becomes congested, the congestion thresholds of AC_VI, AC_BE, and AC_BK decrease by a big margin to reserve the channel access chance for AC_VO with the highest traffic priority. The threshold of AC_VO also declines to satisfy the requirement of transmission collision performance. Actually, the threshold is also affected by the packet generating rate of each AC.

Figure 4.4 shows the performance comparison between the IEEE 802.11p-based EDCA protocol and the dCAP approach in terms of the amount of successful transmissions of each AC. From Fig. 4.4, it can be observed that the two higher priority traffic flows, i.e., AC_VO and AC_VI, have absolutely higher channel

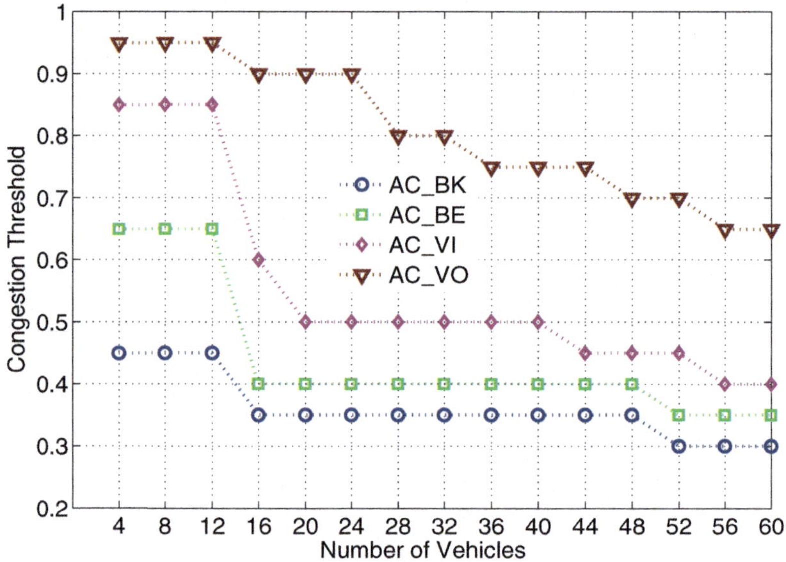

Fig. 4.3 The optimized congestion thresholds over the number of vehicles

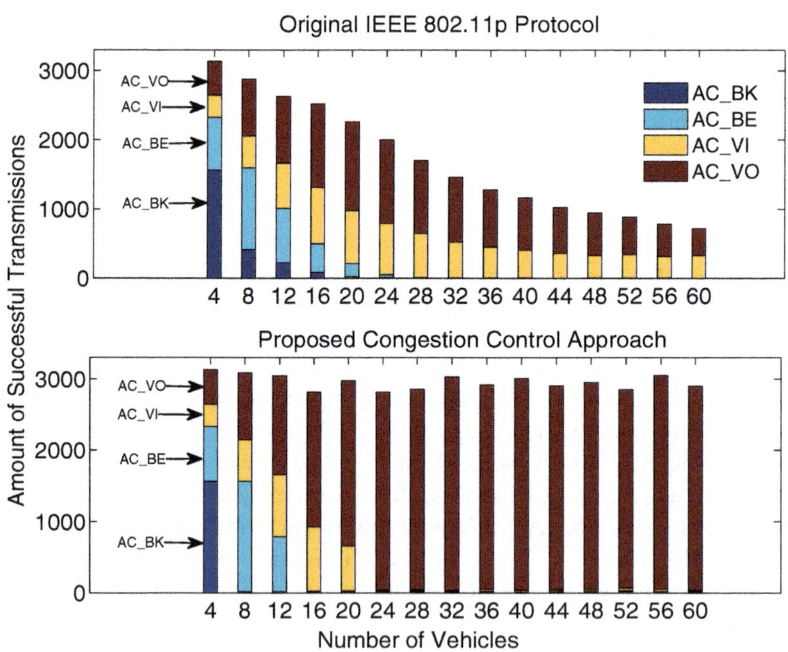

Fig. 4.4 The performance comparison between the dCAP approach and the IEEE 802.11p EDCA protocol in terms of the amount of successful transmissions

accessing chance than the two lower priority traffic flows, i.e., AC_BE and AC_BK, with the EDCA protocol. However, with the EDCA protocol, when there are more vehicles injected into the network, the amount of successful transmissions from AC_VO and AC_VI declines seriously due to the increasing collisions, which occur mainly between the two higher priority traffic flows. While AC_BE and AC_BK seldom have the channel access chance due to the saturated medium in the dense network. In the dCAP approach, since the congestion threshold set for AC_VI is smaller than that set for AC_VO, the AC_VI adjusts the contention window size to reduce the transmission rate ahead of AC_VO in the congested network and hence it reserves more channel access chances for AC_VO. Therefore, the dCAP approach can effectively ease the collision problem between the two higher priority traffic flows and ensure the successful transmission of the highest priority traffic flow AC_VO in the dense vehicular network.

The collision probability of the dCAP approach and the EDCA protocol is demonstrated in Fig. 4.5. It can be found that the with the EDCA protocol, the collisions among the traffic flows are increasing rapidly with the increasing number of vehicles in the network. Whereas with the dCAP approach, the collision probability remains at a low level due to the optimized principle for the congestion thresholds. The result of low transmission collisions would further reduce a large amount of retransmissions.

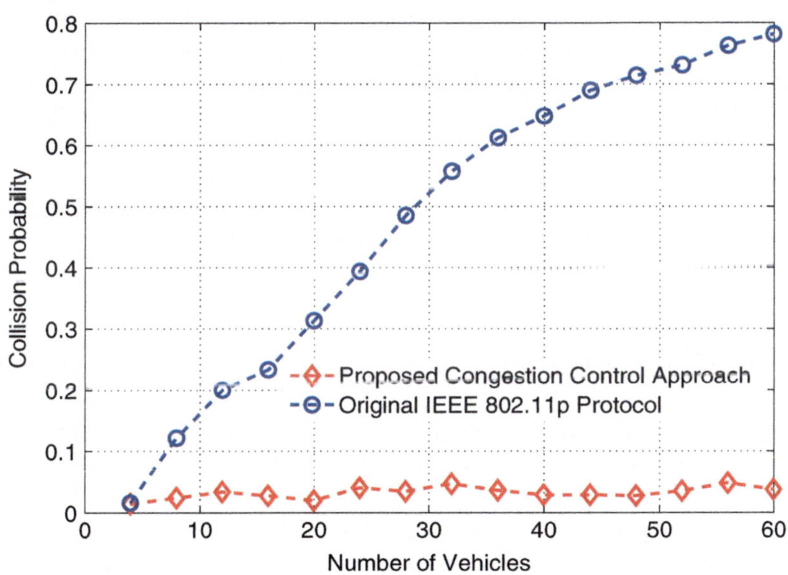

Fig. 4.5 The collision probability comparison between the dCAP control approach and the 802.11p-based EDCA protocol

4.3 Centralized Resource Sharing and Scheduling

Along with the fast development of intelligent vehicles, more rigorous communication requirements for VCN have been raised, such as ultra-high network throughput and ultra-low transmission delay, to support various vehicular applications (e.g., in-vehicle entertainment and autonomous driving). In addition, the architecture of VCN is becoming more complicated with different communication modes, including vehicle-to-vehicle (V2V), vehicle-to-pedestrian (V2P), vehicle-to-infrastructure (V2I), and vehicle-to-cloud (V2C) communications. In future 5G-enabled vehicular networks, all these V2X communications will co-exist and share the same wireless medium for data transmission. V2I communications is between vehicles and a central entity, while V2V and V2P communications are among vehicles and distributed devices. In order to achieve improved network performance under such a complicated and heterogeneous topology, efficient and effective coordination of different types of V2X communications should be provided.

More recently, device-to-device (D2D) communications has been proposed as a novel and enhanced communication mode in cellular networks, in which cellular-controlled short-range direct data transmissions among devices can also access the cellular spectrum [30]. With appropriate resource management, D2D communications can effectively improve the network spectrum efficiency and cell capacity [31–33], increase local data transmission rate, and is an efficient means of cellular data offloading [34]. Due to very similar network topology to the cellular network with D2D communications, the vehicular network is actually naturally feasible for incorporating the concept of D2D communications to seek for joint consideration of different V2X communications and enhanced network performance. Such D2D-based centralized resource sharing and scheduling will significantly improve the network performance of future 5G-enabled VCN. In this section, we first introduce the potential benefits of D2D-enabled VCN, and then discuss more details about D2D for VCN from two perspectives, that is, feasibility and efficiency.

4.3.1 D2D-Enabled VCN

Considering the benefits of D2D communications in cellular networks, the potential gain of introducing D2D concept into VCN can be classified into three types, namely the proximity gain, the hop gain, and the reuse gain. The proximity gain comes from the high data rate and low power consumption due to the relatively short communication range between D2D transceivers. Two nodes (vehicles or pedestrians) in vehicular networks can communicate via a more efficient D2D link through centralized scheduling. This leads to the so-termed hop gain. The reuse gain intuitively comes from the fact that different V2X links can simultaneously share the same radio resources under the D2D underlay mode, which can significantly improve the network spectrum efficiency. Doppler et al. [30] indicated that the overall throughput in the network with D2D communications may increase up to 65% compared to the case where all D2D traffic is transmitted through the traditional cellular mode. Also, the D2D operation can be fully transparent to users and manual pairing or access point definition is not required as in VANETs. In other words, the central controller (e.g., the BS) conceals the complexity of setting up the D2D connections from vehicles or pedestrians.

There are three typical modes for D2D communications:

- Silent mode (no D2D): All available resources are used for cellular links and spectrum reuse is not possible. The D2D devices cannot transmit data and have to stay silent.
- Reuse mode (D2D underlay): D2D devices directly transmit data by reusing some resources of the cellular network. The spectrum reuse can be in either uplink or downlink communications.
- Dedicated mode (D2D overlay): The cellular network dedicates a portion of resources for D2D devices for their direct communications.

The heterogeneous network architecture of vehicular networks provides VCN a natural feasibility to incorporate the D2D concept to enhance the network performance. In [27], the feasibility of employing the D2D concept in vehicular networks by operating the V2V and V2P links as underlaying D2D links was investigated and some guidelines for D2D underlay mode in vehicular network were provided. In order to achieve both effective and efficient D2D-enabled resource sharing among V2I and V2V communications, various interference graph (IG-)based resource sharing schemes were proposed [28, 29], including IG-TDMA and IG-OFDMA, leading to significant improvement in network throughput, in comparison with existing MAC protocols in vehicular networks.

4.3.2 D2D for VCN: Feasibility

In [29], the authors for the first time proposed a D2D-enabled underlaying framework for V2I and V2V communications in vehicular networks, where different V2I and V2V communications are able to share the same resource for data transmissions. Then, in [27], the authors provided an in-depth feasibility analysis of D2D in vehicular networks by operating the V2V/V2P links as underlaying D2D links. Through various simulations of D2D for ITS under vehicular scenarios, four important conclusions have be obtained for D2D for VCN, which can be used as guidelines for future designs exploiting the benefits of D2D to improve the network performance in vehicular networks:

- Spectrum efficiency can be significantly improved with D2D underlay mode.
- There exists an optimum link density for D2D links in the underlay mode, while high density leads to extensive interference and low density results in limited resource reuse.

- Appropriate interference control mechanism should be included to achieve effective resource sharing in D2D underlay mode, especially when the traffic is heavy.
- Due to the fast time-varying fading for V2X communications, channel prediction schemes can enhance the feasibility of D2D under high speed, and position only CSI with prediction has the highest robustness.

4.3.2.1 D2D or Not

The first and foremost question in advocating D2D in VCN is whether such an operation mode would induce any gain, and if so, how much gain. Compared to cellular networks, the vehicular networks result in two major differences from the perspective of D2D communications: (i) the spatial distribution of the vehicles in the vehicular network can be very different from that of the UEs in the cellular network, given the geometry of the road that constrains the vehicle movement; and (ii) the channel characteristics given the much higher mobility of the vehicles in vehicular networks than pedestrians in cellular networks. Channel modeling for the vehicular network is also more challenging than the cellular network because of the rapid change of the vehicular network topology. As a result, although positive results are already well accepted for cellular deployment, the potential gain of D2D in VCN has to be evaluated in realistic vehicular environments.

Considering a single V2V pair with transceiver distance varying from 5 to 100 m, in co-existence of a V2I link, in Figs. 4.6, 4.7, and 4.8, the average spectrum efficiency of four scenarios, namely D2D underlay, D2D overlay, D2D only, and silent (no D2D) modes, for both cellular and vehicular networks, are compared. In all cases, the transmit power is 20 dBm for D2D links and 40 dBm otherwise. The path loss model is $(32.4 + 20log10(d) + 20log10(f))$, where $f = 5.9\,\text{GHz}$ is the carrier frequency. From these figures, it can be observed that the respective curves are very similar when all channels are Rayleigh fading. The vehicular cases start to deviate from the cellular cases when the V2V channels are modeled as Weibull, and the deviation becomes more evident when the V2I channels are also Weibull fading. This implies that the spatial distribution of the vehicles has negligible influence on the spectrum efficiency, and that the more challenging Weibull channel distribution is the determining factor for the spectrum efficiency of all four D2D deployment modes.

Regardless of the channel distribution, however, it can be found that the D2D underlay option always provides the highest spectrum efficiency universally. In addition, when the D2D transceiver distance is less than 15–20 m, the D2D-only mode performs the best and silent mode the worst. This is reversed as the D2D transceivers get further apart. Such is quite reasonable because the D2D link quality is adversely affected by the D2D transceiver distance.

These comparisons demonstrate that: (i) at small D2D transceiver distance, the D2D-only mode facilitates a transmission capacity 2–5 times that of the silent mode, and is thus suitable for high-data-rate communication needs; (ii) at large

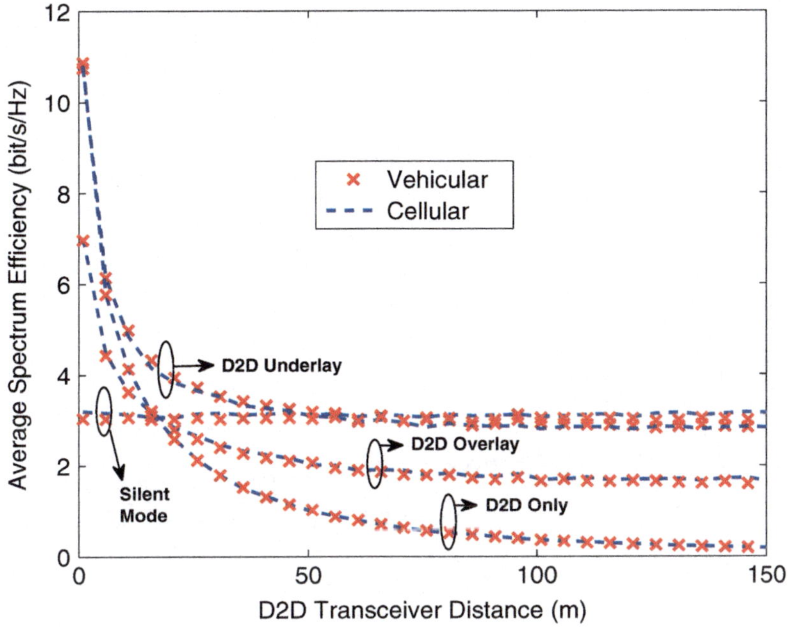

Fig. 4.6 Average spectrum efficiency versus D2D transceiver distance. All channels Rayleigh fading

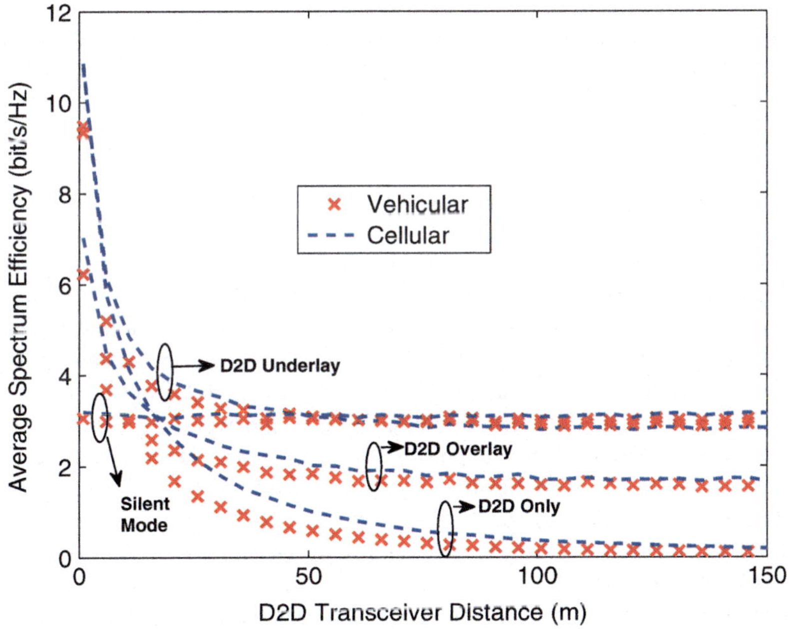

Fig. 4.7 Average spectrum efficiency versus D2D transceiver distance. V2V channels Weibull fading. All other channels Rayleigh fading

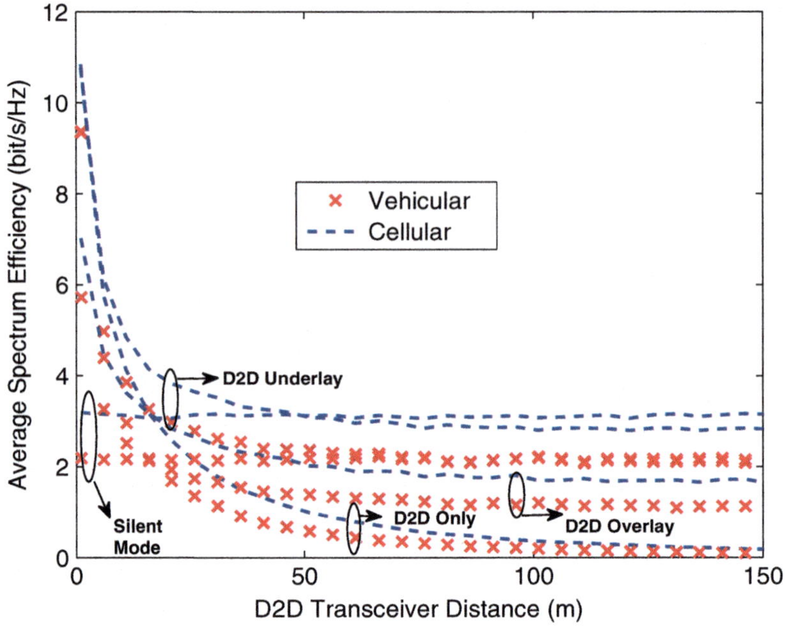

Fig. 4.8 Average spectrum efficiency versus D2D transceiver distance. Cellular channels Rayleigh fading. Vehicular channels Weibull fading

D2D transceiver distances, the silent mode provides a stable data rate capable of supporting safety critical low-data rate applications; and (iii) the D2D-underlay mode strikes the best spectrum efficiency at all D2D transceiver distances.

4.3.2.2 How Many D2D Links

Although the above demonstration clearly shows that the D2D-underlay mode is advantageous in terms of a single-D2D-link spectrum efficiency than all other alternatives, the co-existence of multiple D2D links will introduce interference that may be a detriment to such advantage. To examine this, the scenario with multiple D2D pairs with the D2D link range of 20 m is investigated. From Fig. 4.9, it can be seen that even when all channels are Weibull fading, the average spectrum efficiency (and correspondingly the percentage of active D2D pairs) is noticeably higher for vehicular networks. This implies that, when multiple D2D links coexist, the spatial vehicle distribution has noticeable effects on the interference among these D2D links. In addition, the geometry in vehicular networks appears to be more favorable to the D2D deployment.

From all curves, it can be also observed that the spectrum efficiency first increases and then drops as the number of D2D pairs increases because of the interference among them. At low D2D density, interference is quite limited and the average

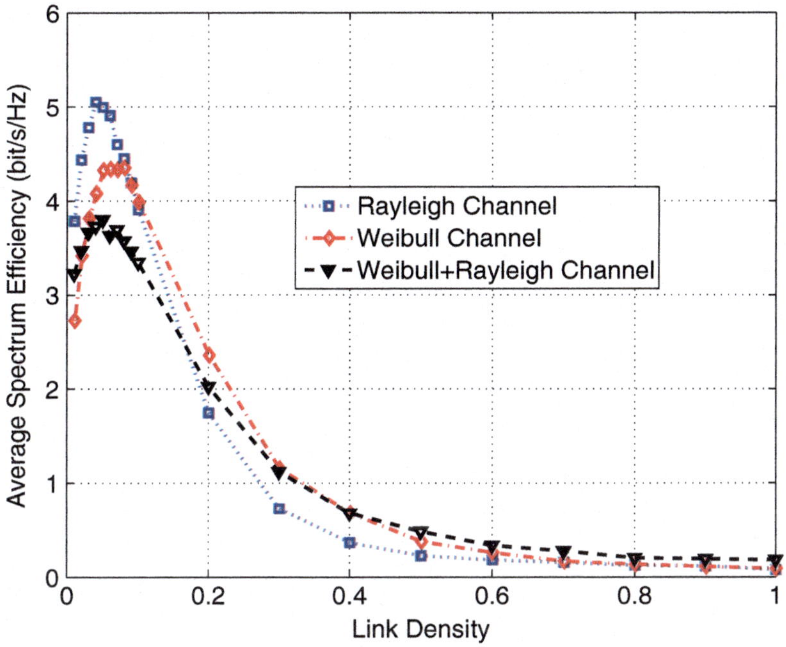

Fig. 4.9 Average spectrum efficiency versus D2D link density, without interference control

spectrum efficiency is mainly determined by the number of D2D links and hence increases with the latter. After a certain point, the interference becomes dominate, hence the average spectrum efficiency decreases as the D2D link density increases. In Weibull channels, the more challenging propagation environment leads to smaller per link capacity, but also smaller interference among different D2D links. As a result, though the average spectrum efficiency in Weibull channels is smaller in the noise-limited range, it is actually the highest in the interference-limited range.

By employing an interference control mechanism that prohibits any D2D pairs to locate within 200 m from each other to share the same resource block, the results shown in Fig. 4.10 indicate that the average spectrum efficiency increases continuously while the percentage of active D2D links remains roughly constant for all cases. Among the different propagation environments, Rayleigh fading channels witness the most prominent performance improvement. Recall that Rayleigh channels also lead to the most significant interference among D2D links, this observation further verifies the effectiveness of such a simple interference control mechanism in eliminating interference.

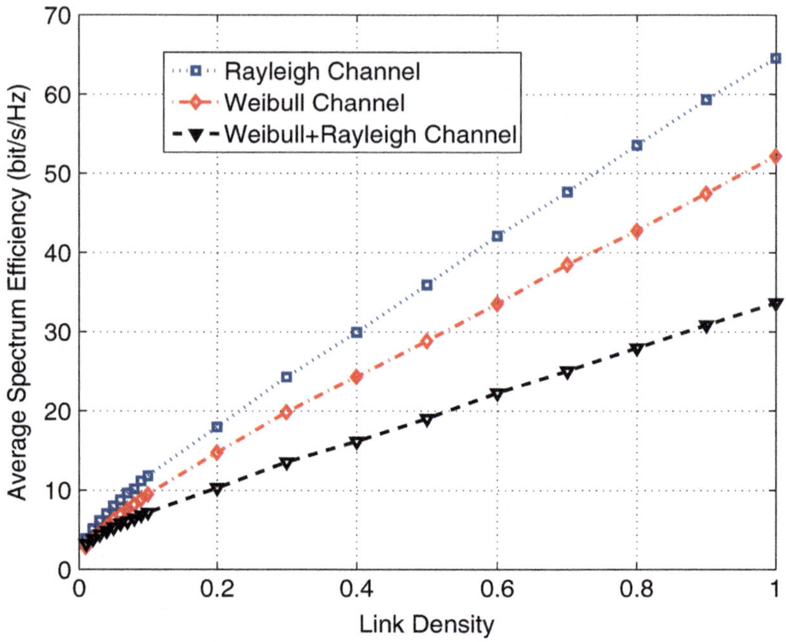

Fig. 4.10 Average spectrum efficiency versus D2D link density, with interference control

4.3.2.3 Full or Partial CSI

Existing resource allocation schemes developed for D2D communications in cellular environments are mostly based on full channel state information (CSI). For scenarios with low device mobility, this may achieve near-optimum performance. For vehicular scenarios with high mobility, however, not only that it is very costly to acquire full CSI, but also the CSI can easily become outdated, especially when it takes non-negligible time to obtain the resource allocation solution.

Hence, the effect of CSI on the D2D-enabled vehicular communications is investigated under two cases, that is, full CSI containing the actual channel fading parameters and partial CSI containing the pass loss only. In Fig. 4.11, we see that as the time cost of information exchange increases, the sum rate of the system drops monotonically for both full and partial CSI cases. However, curves corresponding to these two cases remain very close to each other, demonstrating that the full CSI is no more useful than the vehicle position information when they are both outdated. In other words, due to the rapid channel variations in the mobile scenario, even if one can obtain full CSI, the actual obtained sum rate will still suffer from a significant degradation.

The nice thing is that the position measurements are often readily available in vehicular applications. One can use the position information to obtain the partial CSI, that is, the path loss of the channel based on the transceiver distance only. More

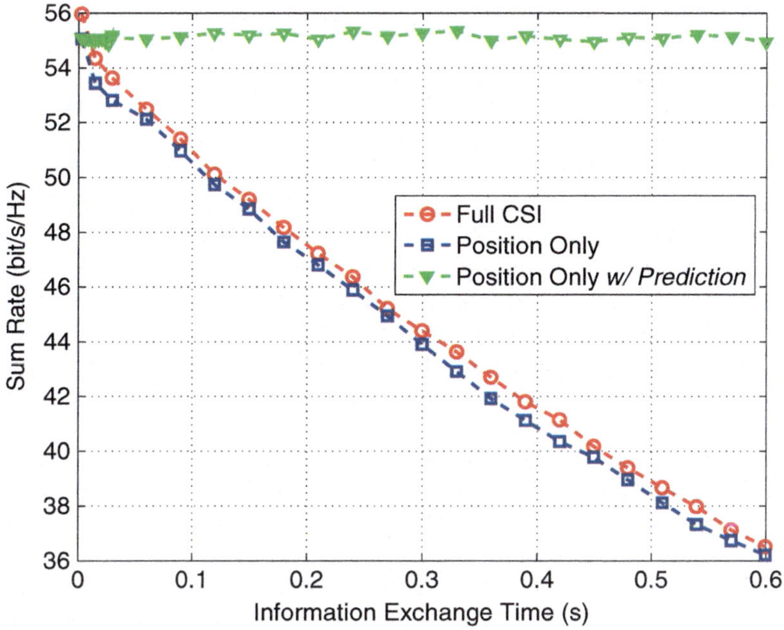

Fig. 4.11 The sum rate resulted from resource allocation with full and partial CSI in Rayleigh fading channels with $\sigma^2 = 5$ dB. Tx power is 46 dBm at RSU and 23 dBm for D2D. The noise power density is −174 dBm/Hz and the bandwidth block size is 15 kHz. Vehicle velocity is 30 ∼ 60 km/h

importantly, the largely predictable vehicle movement makes it possible to estimate the vehicle trajectory, and accordingly the resource allocation algorithm can make use of the predicted partial CSI (i.e., the path loss), at the time of applying the resource allocation solution. The resultant sum rate performance is also plotted in Fig. 4.11. It can be observed that the performance remains almost the same between the real-time position information and the predicted position information. Actually, due to the high complexity of channel estimation in obtaining the full CSI, the partial CSI instead of the full CSI will be more suitable for vehicular networks from the perspective of both practicality and effectiveness.

4.3.3 D2D for VCN: Efficiency

After confirming the feasibility of D2D for VCN in vehicular networks, the next issue is the exploitation of the benefits of D2D communications in vehicular networks for performance improvement. The biggest challenge is the complicated interference scenario caused by flexible resource sharing among V2X communications. In order to achieve both effective and efficient interference management, various interference graphs (IGs) [28, 29, 35] have been proposed in the literature to model the complicated interference scenario in heterogeneous vehicular networks due to resource sharing among different V2X communications, including interference-aware graph (IAG), interference-classified graph (ICG), and interference-free graph (IFG). In all the three IGs, the vertices represent V2I and V2V communication links, while the edges denote the interference between two communication links. Specifically, IAG is constructed based on the full interference information, where the edge weight can be exactly calculated based on the collected CSI. ICG is constructed according to quantized interference information, in which four interference levels are introduced based on the geographic location and the exact CSI of interference links is not required. And IFG is constructed based on binary interference information, which indicates "yes" or "no" for the existence of unacceptable interference subject to a pre-determined interference-free threshold. IAG achieves the highest accuracy of interference modeling, but it requires the complete CSI, which leads to high communication overhead and is not applicable in practical vehicular networks. Nevertheless, ICG and IFG can provide feasible and efficient interference modeling for D2D-enabled vehicular networks. The feature comparisons of IAG, ICG, and IFG are given in Fig. 4.12.

Based on the introduced interference graphs (IGs), two classes of IG-based MAC protocols are provided to achieve both effective and efficient resource sharing among V2I and V2V communications, that is, IG-TDMA protocols [28] and IG-OFDMA protocols [29]. For IG-TDMA protocols, the allocated resource unit is the time slot and the fairness over time, access priority should be taken into consideration for the protocol design. Whereas for IG-OFDMA protocols, the allocated resource unit the resource block (RB) and frequency-selective fading should be considered to distinguish the channel quality of a specific communication link on different RBs.

	IFG	ICG	IAG
Interference Levels	2	4	Exact Value
Interference Accuracy	Low	Medium	High
Exact CSI Required	×	×	✓
Communication Overhead	Low	Low	High
Frequency-Selective Interference	×	×	✓
Link Access Fairness Concern	✓	×	×
Applicable in TDMA	✓	✓	×
Applicable in OFDMA	×	✓	✓

Fig. 4.12 Feature Comparisons of IAG, ICG, and IFG

4.3.3.1 IG-TDMA Protocol

In [28, 36], by utilizing IFG to model the interference scenario in vehicular networks, an efficient IG-TDMA protocol for VCN was proposed. In the centralized IG-TDMA protocol, the RSU, as a centralized controller, collects the channel state information and the individual information of the demanding communication links within its communication coverage and calculates the scheduling weight factor of each communication link, based on which the scheduling decisions will be made.

4.3.3.2 Scheduling Weight Factor Design

The key issue of the centralized IG-TDMA protocol lies in the design of the scheduling weight factor. Each AC of each communication link requesting for data transmission has a corresponding weight factor, denoted by $Q_{k,j}$, $k \in \mathcal{K}$, $j = 1, 2, 3, 4$, which is calculated according to its reported channel state information and individual information. Based on the obtained scheduling weight factors, a scheduling order is decided by the RSU, where AC_j of communication link U_k with a larger scheduling weight factor will be scheduled first to fulfill its transmission demand.

The designed scheduling weight factor mainly consists of three parts, namely the channel quality factor, the speed factor, and the AC factor. The channel quality factor is designed by considering the channel quality of different communication links to optimize the network throughput. The speed factor is provided to achieve the potential serving time fairness among the moving vehicles within various velocities. And the AC factor is set to distinguish different accessing priorities of different ACs.

Then, the scheduling weight factor $Q_{k,j}$, $k \in \mathcal{K}$, $j = 1, 2, 3, 4$, in the IG-TDMA protocol is designed as

$$Q_{k,j} = (CQF_k(t))^{\alpha} \, (SF_k)^{\beta} \left(ACF_j\right)^{\gamma} \tag{4.2}$$

where t denotes the time index of the current transmission frame, and α, β, and γ are the balancing factors. Note that α, β, and γ can be adjusted to practical demands in order to achieve an efficient tradeoff among the effects of the three parts of the designed weight factor.

In order to improve the network throughput of the vehicular networks, the channel quality of different communication links is incorporated into the design of the scheduling weight factor. When maximizing the network throughput, the communication link with the highest channel quality currently is always preferred to be served first. However, in order to guarantee the QoS of the communication links suffering from relative long time bad channel conditions, the fairness among communication links should be also considered. Therefore, similar to the scheduler designed in [37], the channel quality factor of communication link U_k, $k \in \mathcal{K}$, is designed as

$$CQF_k(t) = \frac{C_k(t)}{\mathcal{R}_k(t-1)} \tag{4.3}$$

where $C_k(t) = W\log_2\left(1 + \frac{p_{tr}g_k(t)}{\sigma^2}\right)$ represents the potential transmission rate requested by communication link U_k based on its reported channel state information $g_k(t)$ at the beginning of transmission frame t, p_{tr} denotes the transmit power of the RSU or the c-vehicle depending on whether U_k is a V2I or V2V link (i.e., $p_{tr} = p_r$ if $U_k \in \mathcal{M}$ and $p_{tr} = p_v$ if $U_k \in \mathcal{N}$), and $\mathcal{R}_k(t)$ is the average transmission rate calculated by an iteration process as

$$\mathcal{R}_k(t) = \left(1 - \frac{1}{t_c}\right)\mathcal{R}_k(t-1) + \frac{1}{t_c}R_k(t) \tag{4.4}$$

where t_c is a predefined number of transmission frames denoting the length of the averaging window, and $R_k(t)$ is the total transmission rate acquired by communication link U_k at transmission frame t. The value of the parameter t_c defined here is related to the maximum amount of time for which an individual communication link can be starved (i.e., not receive service for a certain long time). If communication link U_k does not obtain any transmission service from the RSU at transmission frame t, then $R_k(t) = 0$. Note that the average transmission rate is updated by the RSU at the start of each transmission frame for each communication link including the communication links receiving no transmission service during the previous transmission frame.

Due to the mobility pattern in vehicular networks, vehicles with various velocities have quite different residence time within the communication coverage of the RSU/c-vehicles and this leads to a potential serving time fairness problem among the vehicles in vehicular networks. To achieve this potential serving time fairness that the vehicles with different velocities can obtain almost the same possibilities of the transmission service from the RSU/c-vehicles, we design the speed factor based on the accessing probabilities of different vehicles when considering the speed effect only. Therefore, the speed factor of communication link U_k can be designed as

$$SF_k = \left(\left[\frac{L}{\tilde{v}_k T_f}\right]_{\text{int}}\right)^{-1} \tag{4.5}$$

where L is the communication diameter of the RSU or a c-vehicle depending on whether U_k is a V2I or V2V link, T_f is the transmission frame duration, $\tilde{v}_k$ is the absolute value of the relative speed between the receiver and the transmitter of communication link U_k, and $[x]_{\text{int}}$ denotes the largest integer that does not exceed x.

Similar to the EDCA protocol, four different ACs, i.e., AC_j, $j = 1, 2, 3, 4$, are also considered in the IG-TDMA protocol and each has a corresponding priority to indicate its accessing probability. In the EDCA protocol, the priorities of different ACs are distinguished by the predefined minimum and maximum CW values

Table 4.1 Parameters for the four ACs in IEEE 802.11p-based EDCA protocol

AC	AC_1	AC_2	AC_3	AC_4
CW_{min}	3	3	7	15
CW_{max}	7	15	1023	1023

for each AC, which is shown in Table 4.1. In order to guarantee the accessing probabilities of different ACs to stay almost the same with that in the EDCA protocol, the AC factor is designed based on the approximate accessing probabilities of different ACs calculated in the EDCA protocol.

From [38], the accessing probability of a certain AC is approximately inversely proportional to its corresponding minimum CW, which means

$$Pr_{AC}(j) \approx \frac{1}{CW_{min}(\mathrm{AC}_j)} \tag{4.6}$$

where $Pr_{AC}(j)$ denotes the accessing probability of AC_j, $j = 1, 2, 3, 4$. Note that the minimum CWs of AC_1 and AC_2 defined in the EDCA protocol have the same value 3, therefore, in order to distinguish the accessing probability of these two ACs, the maximum CW factor is also employed into the design of the AC factor. Then, the AC factor of AC_j in the IG-TDMA protocol indicating different accessing probabilities of different ACs can be designed as

$$ACF_j = \frac{1}{CW_{min}(\mathrm{AC}_j)} + \frac{1}{CW_{max}(\mathrm{AC}_j)} \tag{4.7}$$

where $CW_{min}(\mathrm{AC}_j)$ and $CW_{max}(\mathrm{AC}_j)$ are given in Table 4.1.

4.3.3.3 IFG-TDMA Protocol

The procedure of the IFG-TDMA protocol is provided in Fig. 4.13. During the scheduling process, the RSU collects the reported information including the current channel state information, the position and relative speed information, and the transmission AC information from the vehicles within its communication coverage, and updates the scheduling weight factors of the ACs of each communication link at the beginning of each transmission frame. Note that any V2V communication links meeting the condition of the resource reusing mode will be merged into an enhanced communication group and the scheduling weight factor of the communication group is the sum of the scheduling weight factors of all the V2V communication links in it. The scheduling weight factors of the communication links/groups are then sorted in the decreasing order. The AC of a communication link with a higher scheduling weight factor will always be prioritized in the current transmission frame until its transmission demand is met. This scheduling process continues until all transmission requests are satisfied or the time slots in the current transmission frame are completely scheduled, then the RSU will broadcast the scheduling decisions

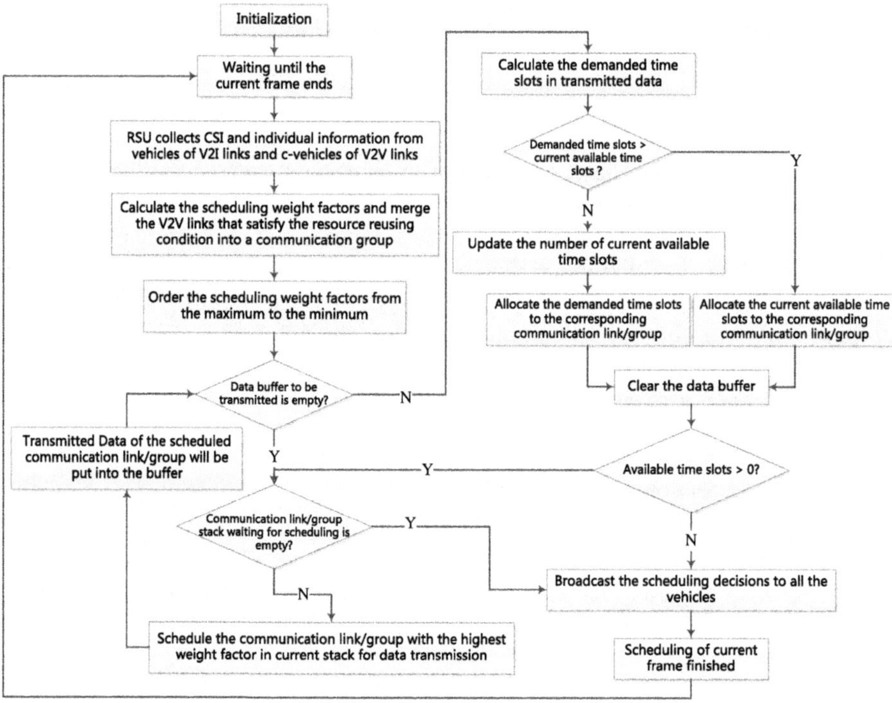

Fig. 4.13 Flow diagram of the IFG-TDMA protocol

to all vehicles, based on which the communication links acquiring resources can perform their individual data transmissions during the allocated time slots.

Figure 4.14 shows the performance comparison in terms of the average spectrum efficiency (defined as the network throughput over unit bandwidth) between the IFG-TDMA protocol and the EDCA protocol in [39]. From Fig. 4.14, it can be seen that the spectrum efficiency achieved with the IFG-TDMA protocol has a significant performance gain compared with that in the EDCA protocol, especially when the number of vehicles in the vehicular network is large. Note that when the number of vehicles in the vehicular network is large, even though the channel quality factor is not incorporated into the scheduling weight factor in the IFG-TDMA protocol, i.e., setting $\alpha = 0$, the achieved performance of the IFG-TDMA protocol is still much better than that of the EDCA protocol. This is because, when the number of vehicles is large, the collisions among different communication links accessing the same time slots are more likely, which leads to significant failure of data transmission and thus noticeable degradation in network performance. Also note that with the IFG-TDMA protocol, the performance under the condition $\alpha = 1, \beta = 0, \gamma = 0$ is always better than that under the condition $\alpha = 1, \beta = 1, \gamma = 1$, however, with the speed factor and the AC factor incorporated into the protocol, it can achieve a much better fairness.

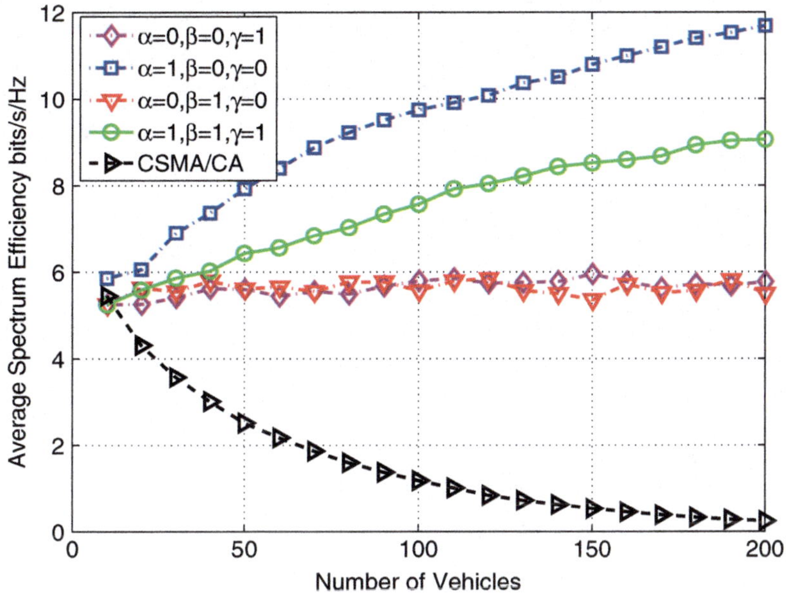

Fig. 4.14 Average spectrum efficiency performance comparison between the IFG-TDMA protocol and the EDCA protocol

Figure 4.15 shows the fairness coefficients in the IFG-TDMA protocol as well as the EDCA protocol. Note that according to the definition of fairness coefficient in [38], a value closer 1 means a more fair situation, while 1 indicates absolute fairness among all communication links. From Fig. 4.15, it can be seen that the fairness coefficient with all the three weight factors (i.e., the channel quality factor, the speed factor, and the AC factor) in the IFG-TDMA protocol is closer 1 than those without any of the weight factors as well as that with the EDCA protocol, which verifies the efficiency of the designed weight factors in the IFG-TDMA protocol in solving the fairness problem in vehicular networks.

4.3.3.4 IG-OFDMA Protocol

In [29], based on the IAG and ICG, two IG-OFDMA protocols (i.e., IAG-OFDMA and ICG-OFDMA) were proposed to achieve both effective and efficient resource sharing among different V2I and V2V links, with the objective of optimizing the vehicular network throughput.

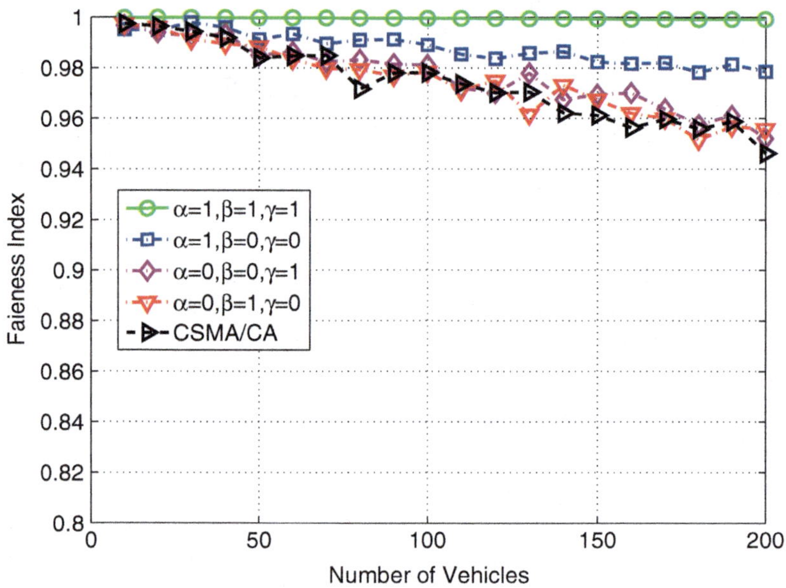

Fig. 4.15 Fairness coefficient comparison

4.3.3.5 IAG-OFDMA

To provide the IAG-OFDMA protocol, several definitions are required as follows.

Definition 4.1 The *cluster value* of RB_k's cluster $\mathcal{C}_k$, denoted by $v_c(\mathcal{C}_k)$, is defined as the sum of the channel capacity for all the communication links belonging to the cluster taking the mutual interference among them into consideration. Therefore, the cluster value $v_c(\mathcal{C}_k)$ can be given as

$$v_c(\mathcal{C}_k) = \sum_{V_i \in \mathcal{C}_k} \frac{W}{K} \log_2 \left(1 + \text{SINR}_{V_i}^k\right) \tag{4.8}$$

where $\text{SINR}_{V_i}^k$ is the SINR of the communication link V_i over RB_k.

Definition 4.2 The *interference value* of RB_k's cluster $\mathcal{C}_k$, denoted by $v_i(\mathcal{C}_k)$, is defined as the sum of the mutual interference between every two communication links belonging to the cluster. Therefore, the interference value $v_i(\mathcal{C}_k)$ can be given as

$$\begin{aligned} v_i(\mathcal{C}_k) &= \sum_{V_i, V_j \in \mathcal{C}_k, V_i \neq V_j} (I_{V_i,V_j}^k + I_{V_j,V_i}^k) \\ &= \sum_{V_i, V_j \in \mathcal{C}_k, V_i \neq V_j} E_{V_i,V_j}^k \end{aligned} \tag{4.9}$$

where $E^k_{V_i,V_j}$ can be easily obtained from the edge weight of the IAG.

Definition 4.3 The *virtual cluster* of RB_k, denoted by $\mathcal{C}^\star_k$, is defined as the set of the vertices whose current interested RB index is k, i.e., $\mathcal{C}^\star_k = \{V_i \mid \delta(V_i) = k,\ V_i \in \mathcal{V}\}$.

The basic idea of the IAG protocol is to iteratively gather vertices from the virtual clusters into the corresponding clusters of the same RB, taking both the interference value and the cluster value into account to guarantee that the cluster value of each cluster is maximized.

As demonstrated in Table 4.2, at the start of the resource assignment process, the IAG that indicates the current situation of the vehicular network is constructed based on the channel and identity information collected from the V2I and V2V communication links by the roadside infrastructure. The parameters of the interference-aware graph are initialized. Considering the cluster $\mathcal{C}_k$ of RB_k, a vertex V^* is first selected from the virtual cluster $\mathcal{C}^\star_k$ to satisfy $V^* = \arg\min_{V_i \in \mathcal{C}^\star_k} v_i\left(\mathcal{C}_k \bigcup \{V_i\}\right)$. Note that if $\mathcal{C}_k = \Phi$, the selected vertex should be the one that can achieve the largest cluster value in $\mathcal{C}_k$, i.e., $V^* = \arg\max_{V_i \in \mathcal{C}^\star_k} v_c\left(\mathcal{C}_k \bigcup \{V_i\}\right)$. By comparing the cluster value $v_c(\mathcal{C}_k)$ with $v_c\left(\mathcal{C}_k \bigcup \{V^*\}\right)$, if $v_c\left(\mathcal{C}_k \bigcup \{V^*\}\right) > v_c(\mathcal{C}_k)$, one adds V^* into $\mathcal{C}_k$, i.e., $\mathcal{C}_k = \mathcal{C}_k \bigcup \{V^*\}$, and set $\tau(V^*) = k$ and $\delta(V^*) = 0$, otherwise deletes the first index in $\mathcal{L}(V^*)$ and update $\delta(V^*)$ as well as all the virtual clusters. Note that whenever the individual information of a vertex in the IAG changes, the virtual clusters should be updated correspondingly. This iterative RB assignment process cycles until all the virtual clusters are empty, i.e., $\mathcal{C}^\star_k = \Phi$ with $k \in \mathcal{K}$. Finally, the RB assignment solution is composed of $\{\tau(V_i) \mid V_i \in \mathcal{V}\}$. The roadside infrastructure announces

Table 4.2 IAG-OFDMA Protocol

1. Construct the IAG and initialize the parameters of the IAG.
   ⋆ Calculate each edge weight $\mathbf{E}_{V_i,V_j}$ with $V_i, V_j \in \mathcal{V}$ and $V_i \neq V_j$;
   ⋆ Initialize each vertex's individual information, i.e., $\mathcal{L}(V_i)$ and $\delta(V_i)$, and set $\tau(V_i) = 0$ with $V_i \in \mathcal{V}$;
   ⋆ Set $\mathcal{C}_k = \Phi$, where Φ represents an empty set, and $v_i(\mathcal{C}_k) = v_c(\mathcal{C}_k) = 0$ with $k \in \mathcal{K}$;
   ⋆ Initialize $\mathcal{C}^\star_k = \{V_i \mid \delta(V_i) = k,\ V_i \in \mathcal{V}\}$ with $k \in \mathcal{K}$.
2. **Repeat**
   ⋆ Select a vertex V^* from the virtual cluster $\mathcal{C}^\star_k$ with $k \in \mathcal{K}$:
   - If $\mathcal{C}_k = \Phi$, then $V^* = \arg\max_{V_i \in \mathcal{C}^\star_k} v_c\left(\mathcal{C}_k \bigcup \{V_i\}\right)$;
   - Else, $V^* = \arg\min_{V_i \in \mathcal{C}^\star_k} v_i\left(\mathcal{C}_k \bigcup \{V_i\}\right)$.

   ⋆ Compare the cluster value $v_c(\mathcal{C}_k)$ with $v_c\left(\mathcal{C}_k \bigcup \{V^*\}\right)$:
   - If $v_c\left(\mathcal{C}_k \bigcup \{V^*\}\right) > v_c(\mathcal{C}_k)$, then $\mathcal{C}_k = \mathcal{C}_k \bigcup \{V^*\}$, $\tau(V^*) = k$, and $\delta(V^*) = 0$;
   - Else, delete the first index in $\mathcal{L}(V^*)$ and update $\delta(V^*)$.

   ⋆ Update all the virtual clusters.
   Until $\mathcal{C}^\star_1 \bigcup \mathcal{C}^\star_2 \bigcup \cdots \bigcup \mathcal{C}^\star_K = \Phi$.
3. The RB assignment solution is composed of $\{\tau(V_i) \mid V_i \in \mathcal{V}\}$.

the RB assignment solution to the corresponding communication links, then the V2I and V2V communication links can perform their individual data transmission on the allocated RBs.

4.3.3.6 ICG-OFDMA

Different from the edge weight in the IAG that can be calculated accurately by the collected interference channel state information, in the ICG, each edge weight is classified, denoted by W_{V_i,V_j} with $V_i, V_j \in \mathcal{V}$ and $V_i \neq V_j$, into four levels, i.e., no interference, medium interference, significant interference, and infinity interference. The four interference levels between two vertices in the ICG are classified based on the relative geographic location and the identity information of the considered two communication links, and can be defined as follows.

- *No Interference*: Neither receiver of the two communication links is located within the interference range of the transmitter of the other communication link, i.e., the cases (a) and (e) shown in Fig. 4.16.
- *Medium Interference*: Only one receiver of the two communication links is located within the interference range of the transmitter of the other communication link, i.e., the cases (b), (c), (f), and (g) shown in Fig. 4.16.
- *Significant Interference*: Both the receivers of the two communication links are both located within the interference range of the transmitter of the other communication link and at least one of the two communication links is V2V communication link, i.e., the cases (d) and (h) shown in Fig. 4.16.
- *Infinity Interference*: If both the considered communication links are V2I links, the interference between them should be set infinity or equal to a sufficiently large value to guarantee our assumption that resource sharing among different V2I communication links are forbidden.

The aforementioned analysis is performed for every pair of the vertices in the graph to determine the interference level of each corresponding edge. There are only four possible weight values of each edge in the ICG, i.e., W_0, W_1, W_2, and W_n, corresponding to no interference, medium interference, significant interference, and infinity interference, respectively. Besides, the four weight values are ranked as $W_0 \ll W_1 < W_2 \ll W_n$, which indicates different degrees of the interference. Therefore, the edge weight W_{V_i,V_j} with $V_i, V_j \in \mathcal{V}$ and $V_i \neq V_j$ can be given as

$$W_{V_i,V_j} = \begin{cases} W_0, & \text{If } V_i \notin \mathcal{V}_j^{neighbor}, V_j \notin \mathcal{V}_i^{neighbor} \\ W_1, & \text{If } V_i \notin \mathcal{V}_j^{neighbor}, V_j \in \mathcal{V}_i^{neighbor} \\ & \quad \text{or } V_i \in \mathcal{V}_j^{neighbor}, V_j \notin \mathcal{V}_i^{neighbor} \\ W_2, & \text{If } V_i \in \mathcal{V}_j^{neighbor}, V_j \in \mathcal{V}_i^{neighbor} \\ & \quad \text{and } V_i \in \mathcal{B} \text{ or } V_j \in \mathcal{B} \\ W_n, & \text{If } V_i \in \mathcal{A} \text{ and } V_j \in \mathcal{A} \end{cases} \tag{4.10}$$

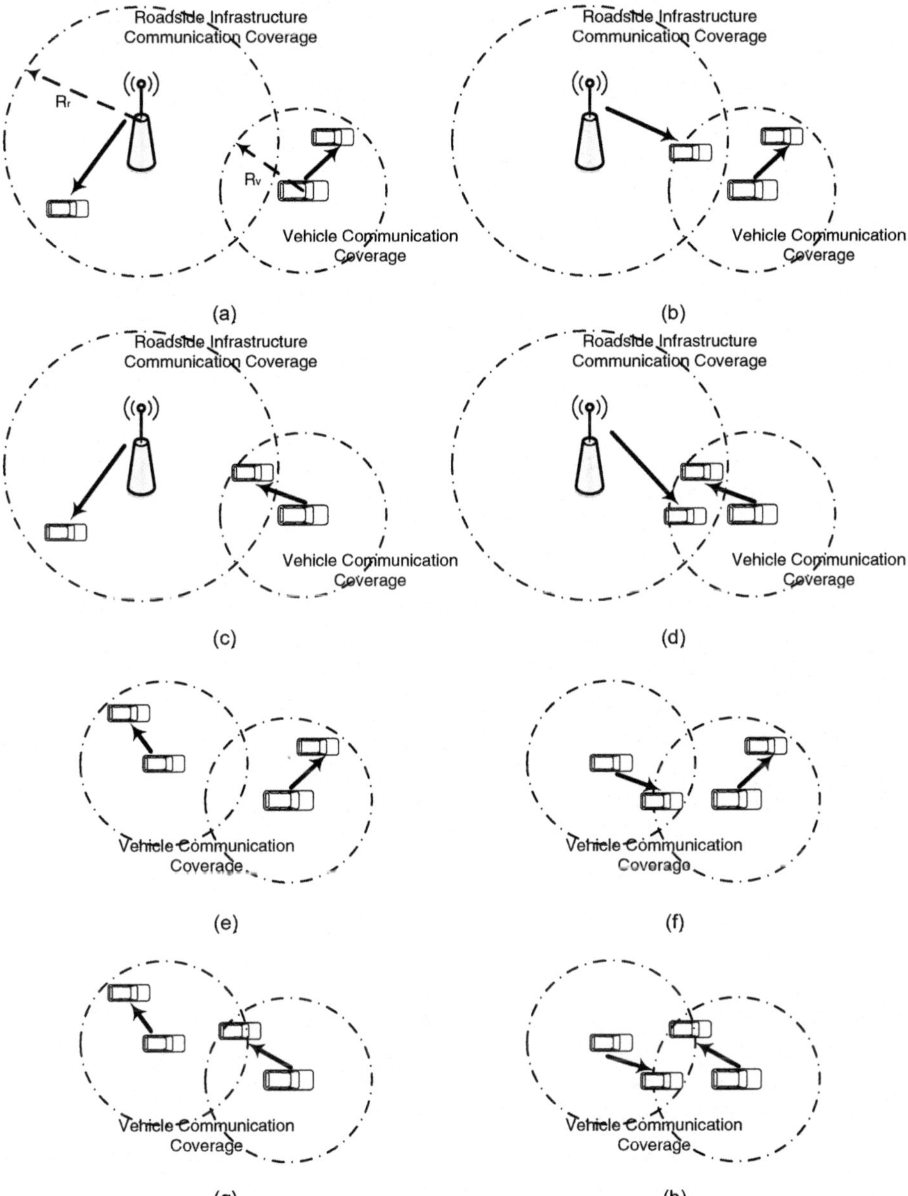

Fig. 4.16 Different cases for relative geographic location of two different communication links

Note that the mutual interference in the ICG, i.e., the edge weight, cannot distinguish different channel responses on different RBs. Therefore, we assume the mutual interference between two vertices V_i and V_j on the RB$_k$, denoted by $W^k_{V_i,V_j}$, is an equal division of the corresponding edge weight, i.e., $W^k_{V_i,V_j} = \frac{W_{V_i,V_j}}{K}$ with $V_i, V_j \in \mathcal{V}$, $V_i \neq V_j$, and $k \in \mathcal{K}$. Therefore, the interference value in the ICG is given as

$$\begin{aligned} v_i(\mathcal{C}_k) &= \sum_{V_i, V_j \in \mathcal{C}_k, V_i \neq V_j} W^k_{V_i,V_j} \\ &= \sum_{V_i, V_j \in \mathcal{C}_k, V_i \neq V_j} \frac{W_{V_i,V_j}}{K}. \end{aligned} \tag{4.11}$$

In Table 4.3, the procedures of the ICG-OFDMA protocol are provided. After the construction and initialization of the ICG, the iterative resource assignment process starts. Considering the cluster $\mathcal{C}_k$ of the RB$_k$, we first select a vertex V^* from the virtual cluster $\mathcal{C}^\star_k$ to satisfy $V^* = \arg \min_{V_i \in \mathcal{C}^\star_k} v_i\left(\mathcal{C}_k \bigcup \{V_i\}\right)$. Note that there is no cluster value concept in this algorithm due to the lack of the accurate interference channel information. Therefore, the number of vertices in one cluster are limited by a threshold T_{ic} to avoid too much vertices joining in one cluster, which may degrade the system performance due to the increasing interference when the number of a vertices in a cluster grows up. If $\mid \mathcal{C}_k \mid < T_{ic}$, where $\mid \mathcal{C}_k \mid$ represents the number of vertices in $\mathcal{C}_k$ at present, one adds V^* into $\mathcal{C}_k$, i.e., $\mathcal{C}_k = \mathcal{C}_k \bigcup \{V^*\}$, and set $\tau(V^*) = k$ and $\delta(V^*) = 0$. Note that the RB index k will be deleted from all the current RB index vectors $\mathcal{L}(V_i)$ with $V_i \in \mathcal{V}$, $V_i \notin \mathcal{C}_k$,

Table 4.3 ICG-OFDMA Protocol

1. Construct the ICG and initialize the parameters of the ICG.
   ⋆ Calculate each edge weight W_{V_i,V_j} with $V_i, V_j \in \mathcal{V}$ and $V_i \neq V_j$;
   ⋆ Initialize each vertex's individual information, i.e., $\mathcal{L}(V_i)$ and $\delta(V_i)$, and set $\tau(V_i) = 0$ with $V_i \in \mathcal{V}$;
   ⋆ Set $\mathcal{C}_k = \Phi$, where Φ represents an empty set, and $v_i(\mathcal{C}_k) = 0$ with $k \in \mathcal{K}$;
   ⋆ Initialize $\mathcal{C}^\star_k = \{V_i \mid \delta(V_i) = k, V_i \in \mathcal{V}\}$ with $k \in \mathcal{K}$.
2. **Repeat**
   ⋆ Select a vertex $V^* = \arg \min_{V_i \in \mathcal{C}^\star_k} v_i\left(\mathcal{C}_k \bigcup \{V_i\}\right)$ from the virtual cluster $\mathcal{C}^\star_k$ with $k \in \mathcal{K}$:
   • If $\mid \mathcal{C}_k \mid < T_{ic}$, then $\mathcal{C}_k = \mathcal{C}_k \bigcup \{V^*\}$, $\tau(V^*) = k$, and $\delta(V^*) = 0$;
   • Else, delete the RB index k from all the current RB index arrays $\mathcal{L}(V_i)$, $V_i \in \mathcal{V}$ and $V_i \notin \mathcal{C}_k$, $k \in \mathcal{K}$.
   ⋆ Update all the virtual clusters.
   Until $\mathcal{C}^\star_1 \bigcup \mathcal{C}^\star_2 \bigcup \cdots \bigcup \mathcal{C}^\star_K = \Phi$.
3. The RB assignment solution is composed of $\{\tau(V_i) \mid V_i \in \mathcal{V}\}$.

and $k \in \mathcal{K}$ once $\mid \mathcal{C}_k \mid = T_{ic}$. Whenever the individual information of a vertex in the ICG changes, the virtual clusters should be updated correspondingly. Also this iterative RB assignment process cycles until all the virtual clusters are empty, i.e., $\mathcal{C}_k^{\star} = \Phi$ with $k \in \mathcal{K}$. Finally, the RB assignment solution is composed of $\{\tau(V_i) \mid V_i \in \mathcal{V}\}$. The roadside infrastructure announces the RB assignment solution to the corresponding communication links, then the V2I and V2V communication links can perform their individual data transmission on the allocated RBs.

In [29], the network sum-rate performance comparison of IAG-OFDMA, ICG-OFDMA, and the traditional greedy OFDMA in vehicular networks was evaluated, in which the results shown in Fig. 4.17 indicate that both IAG-OFDMA and ICG-OFDMA achieve much improved network performance compared with the traditional greedy OFDMA applied in V2X communications. This verifies the efficiency of D2D-enabled underlay resource sharing framework in vehicular networks with various V2X communications. In addition, it is obvious that the IAG-OFDMA protocol achieves better performance than the ICG-OFDMA protocol due to the accurate interference information of the AG. However, the IAG-OFDMA protocol has much higher communication overhead than the ICG-OFDMA protocol. Therefore, the ICG-OFDMA protocol is easier to be implemented into practical vehicular networks.

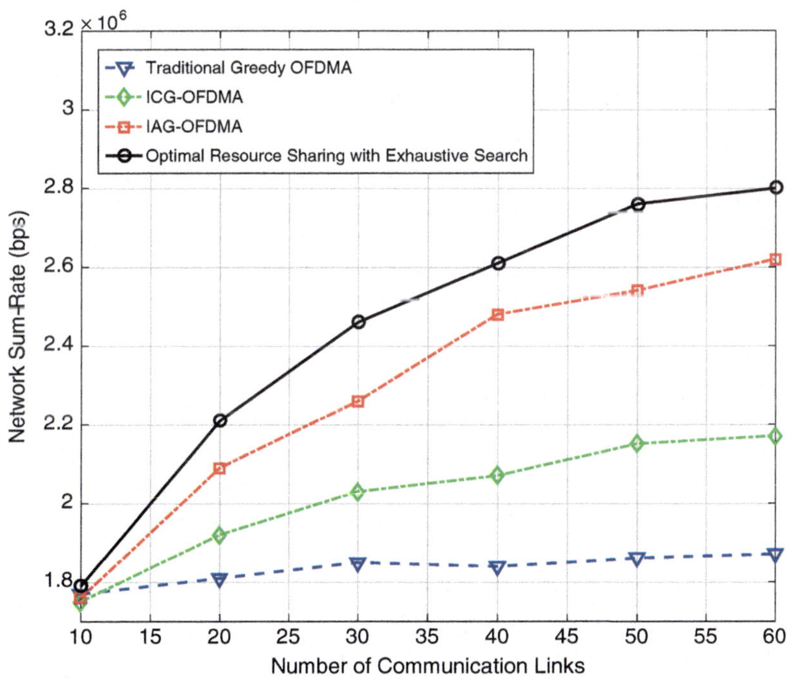

Fig. 4.17 Network sum-rate performance comparison for IAG-OFDMA, ICG-OFDMA, and traditional greedy OFDMA in vehicular networks

4.4 Centralized Data Dissemination

Data dissemination has been regarded as a promising means towards efficient big data/file transmissions among moving vehicles and become an essential and attractive problem in vehicular networks. However, due to the high mobility and heterogeneous network feature of vehicular networks, there are many challenges in achieving efficient data dissemination, including fast-changing topology, limited connection time for V2X links, and severe collision problem when the traffic is heavy. In the literature, most works focused on network coding (NC)-based data dissemination approaches, mainly including random linear NC (RLNC)-based approaches [40, 41] and fountain code/rateless code (FC/RC)-based approaches [42–44], which can reduce duplicate transmissions and simplify the transmission scheduling. However, these NC-based approaches are mostly distributed ones, which may suffer from severe collision problem when the traffic in the vehicular network becomes heavy and cannot achieve a joint global optimization by taking both V2I and V2V communications into consideration. In [45], a novel centralized data dissemination scheduling was proposed by introducing the space-time network coding (STNC) [46] that can effectively exploit the diversity gain to improve data transmission reliability. In this section, we discuss the centralized data dissemination scheduling in details.

4.4.1 Centralized Data Dissemination Scheduling

In the centralized data dissemination scheduling protocol proposed in [45], a central server (CS) connected with RSUs via wired backhaul makes the scheduling decision based on the collected CSI and vehicle information, which can effectively avoid data transmission collisions at the area edge due to non-cooperative RSUs, as illustrated in Fig. 4.18a, b.

4.4.1.1 Space-Time Network Coding (STNC)

Networking coding is usually applied for data broadcasting without acknowledgment since it can effectively reduce redundant transmissions and simplify the transmission scheduling [47, 48]. However, due to the complex decoding issues of the traditional NC tools, i.e., RLNC and Fountain Codes, the proposed centralized data dissemination scheduling exploits a simple and new NC method, namely STNC. Assume that there are N packets, i.e., $\mathbf{X} = \{X_n | n = 1, 2, \ldots, N\}$ to be disseminated from the RSU to the vehicles in the area of interest (AoI) and each packet contains Q symbols. A single packet X_C is generated by combining the N packets with the STNC method in the following manner

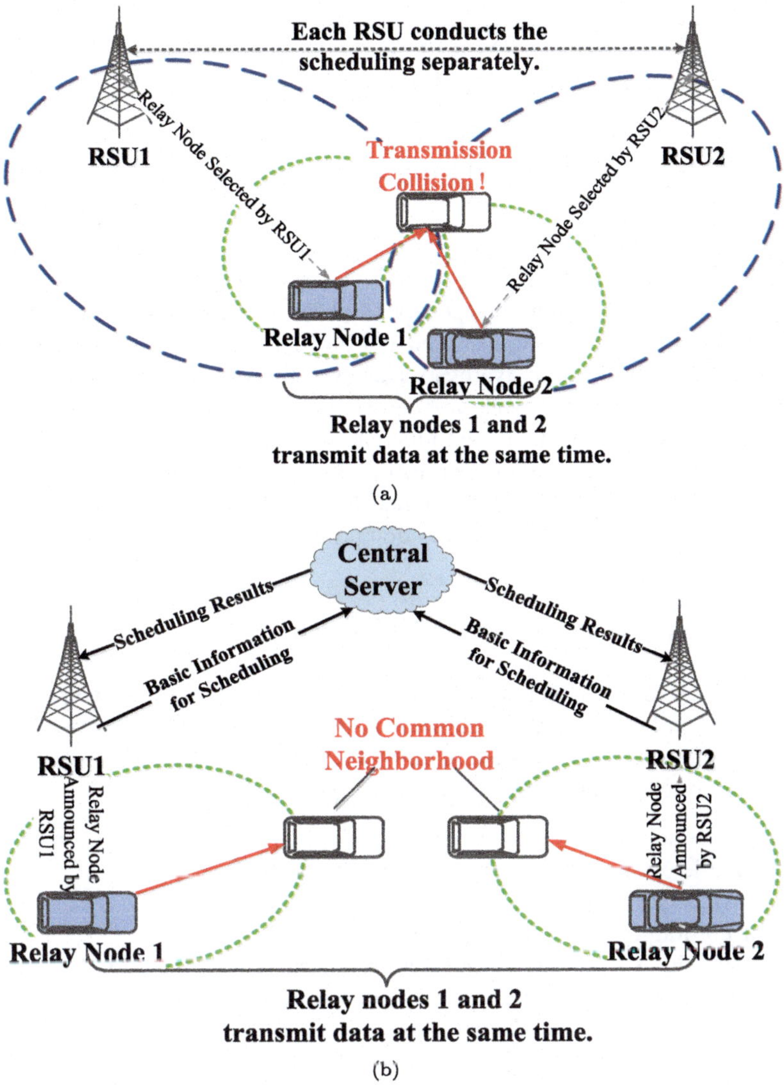

Fig. 4.18 (**a**) The collision illustration when the RSUs select relay nodes separately. (**b**) The collision avoidance with the CS

$$X_C^i(t) = \sum_{n=1}^{N} X_n^i \alpha_n(t), \quad i = 1, 2, \ldots, Q \tag{4.12}$$

where X_C^i and X_n^i are the ith symbol of packets X_C and X_n, respectively, and $\alpha_n(t)$ is the coding coefficient in the form of a complex-valued signature waveform to protect X_n against the interference from other symbols. The cross correlation between the coding factor is $\rho_{mn} = \langle \alpha_m(t), \alpha_n(t) \rangle$, where $\langle f(t), g(t) \rangle \triangleq \frac{1}{T_s} \int_0^{T_s} f(t)\, g^*(t) dt$, and $\rho_{nn} = 1$. For the most investigated RLNC method, the coding coefficients are randomly generated at the intermediate nodes. Therefore, the coding factors should be transmitted along with the corresponding coded packet which results in high system overhead. However, in the STNC method, the coding factors remain the same for each intermediate node and thus can be broadcast only once from the original data source to the end receivers. Therefore, compared with the RLNC method, the system overhead gets significantly reduced in the STNC method. Moreover, with STNC the signal can be detected by a simple matched filtering method considering the correlation property among coding coefficients.

4.4.1.2 Data Dissemination Scheduling Design

The centralized data dissemination scheduling by the CS is performed periodically and the interval between two successive scheduling intervals is denoted as a data dissemination cycle. As illustrated in Fig. 4.19, the procedure of the centralized data dissemination scheduling contains three main phases:

1. Relay selection phase: The CS makes centralized scheduling decision by selecting suitable relay nodes for each transmission frame.
2. Relay transmission phase: All the nodes including vehicles and RSUs broadcast data according to the scheduling results.
3. Feedback phase: Each vehicle updates its current velocity, position, and decoding status to its connected RSU for the scheduling to the next data dissemination cycle.

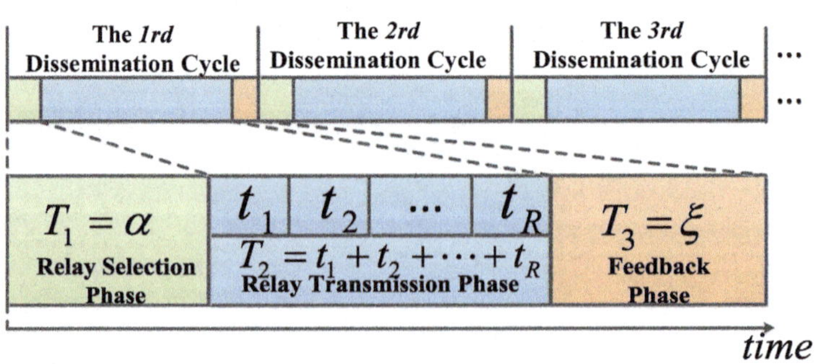

Fig. 4.19 The framework of the centralized data dissemination

4.4.1.3 Relay Selection Phase

For the first data dissemination cycle, the relay selection is based on the initial data decoding status, i.e., no vehicles has received any data. While for the later data dissemination cycle, the relay selection is based on the decoding status reported by the vehicles at the end of former cycle. Since the RSU is the original source, the relay selection strategy should guarantee that RSUs are the selected nodes for the first transmission frame t_1 in the first data dissemination cycle.

The aim of the relay selection strategy is to assign the current transmission frame to the node set with the maximum dissemination utility. When the packet decoding state at a vehicle changes from failure to success after receiving the forwarding data from a relay, we then consider the relay as useful. The designed strategy generates the selected node set Ω_1 first and Ω_R last according to the transmission frame order. For each transmission frame t_r for $r = 1, 2, \ldots, R$, the required information to run the relay selection includes the predicted decoding status after the transmission frame t_{r-1} and vehicle velocity and position. Specifically, the decoding status of packet X_n at node R_x ($R_x \in \{\mathbf{RSU} \cup \mathbf{U}\}$) predicted after t_{r-1} is denoted as $\tilde{\beta}^{t_{r-1}}_{R_x n}$. For the first transmission frame, i.e, $r = 1$, $\tilde{\beta}^{t_0}_{R_x n}$ is equal to the initial decoding status at the start of the dissemination cycle. For the frame t_r for $r = 2, 3, \ldots, R$, $\tilde{\beta}^{t_{r-1}}_{R_x n}$ is predicted based on the relay selection result Ω_{r-1}. Note that if R_x is an RSU, $\tilde{\beta}^{t_{r-1}}_{R_x n} \equiv 1$. Moreover, if R_x is a vehicle, for the first dissemination cycle $\tilde{\beta}^{t_0}_{R_x n} = 0$ and for the later dissemination cycle $\tilde{\beta}^{t_0}_{R_x n}$ is equal to the decoding status reported by vehicles at the end of the corresponding former dissemination cycle.

Given that $R_x (R_x \in \{\mathbf{RSU} \cup \mathbf{U}\})$ is the candidate relay node. The relay selection process for each transmission frame t_r for $r = 1, 2, \ldots, R$ is carried out as follows:

Step 1 [SNR Calculation]: To obtain the node utility, the SNR is first calculated for packet decoding. With MRC, the SNR at U_k to decode the packet X_n after data relaying from R_x is

$$\gamma^{t_r, R_x}_{kn} = \sum_{i=1}^{r-1} \frac{P^i_k L^i_k \left\| h^i_k \right\|^2}{N_0 \varepsilon_n} \tilde{\beta}^{t_{i-1}}_{ikn} + \frac{P_{R_x} L^{t_r}_{R_x k} \left\| h^{t_r}_{R_x k} \right\|^2}{N_0 \varepsilon_n} \tilde{\beta}^{t_{r-1}}_{R_x n} \tag{4.13}$$

where P_{R_x} is the transmit power of R_x and $L^{t_r}_{R_x k}$ is the path loss between R_x and U_k when R_x disseminates data in frame t_r.

Furthermore, $L^{t_r}_{R_x k}$ is dependent on the distance between R_x and U_k and is given by $L^{t_r}_{R_x k} = f_L\left(d^{t_r}_{R_x k}\right)$, where f_L is the distance dependent path loss and $d^{t_r}_{R_x k} = \sqrt{\left(X^{t_r}_{R_x} - X^{t_r}_k\right)^2 + \left(Y^{t_r}_{R_x} - Y^{t_r}_k\right)^2}$ is the distance between R_x and U_k in the frame t_r. $\left(X^{t_r}_{R_x}, Y^{t_r}_{R_x}\right)$ and $\left(X^{t_r}_{U_k}, Y^{t_r}_{U_k}\right)$ denote the position of R_x and U_k in the frame t_r via XY axes, respectively. We then calculate the position of node $U \in \{R_x, U_k\}$ by $X^{t_r}_U = X^{t_f}_U + v^X_U \cdot T_{rf}$ and $Y^{t_r}_U = Y^{t_f}_U + v^Y_U \cdot T_{rf}$, where $\left(X^{t_f}_U, Y^{t_f}_U\right)$ denotes the feedback position from node U at time t_f, $\boldsymbol{v}_U = \left(v_U^X, v_U^Y\right)$ is the corresponding feedback XY-velocity components of node U, and T_{rf} is the duration between the frame t_r and the feedback time t_f.

Moreover, the corresponding average SNR is formulated as

$$\bar{\gamma}_{kn}^{t_r,R_x}=E\left[\gamma_{kn}^{t_r,R_x}\right]=\sum_{i=1}^{r-1}\frac{P_k^i L_k^i \sigma_h^2}{N_0\varepsilon_n}\tilde{\beta}_{ikn}^{t_{i-1}}+\frac{P_{R_x} L_{R_x k}^{t_r}\sigma_h^2}{N_0\varepsilon_n}\tilde{\beta}_{R_x n}^{t_r}. \tag{4.14}$$

Step 2 [Utility Calculation]: After receiving the signal relayed by R_x in the transmission frame t_r, the decoding status to decode X_n at U_k is given by

$$\tilde{\beta}_{kn}^{t_r,R_x} = \begin{cases} 1, & \bar{\gamma}_{kn}^{t_r,R_x} \geq \gamma_{th} \\ 0, & \bar{\gamma}_{kn}^{t_r,R_x} < \gamma_{th} \end{cases} \tag{4.15}$$

where γ_{th} denotes the successful decoding threshold. Therefore, the utility of R_x as a relay node for the transmission frame t_r is calculated by

$$\Phi^{t_r,R_x}= \sum_{k=1,\mathrm{U}_k\in\mathcal{N}_{R_x}^r}^{K} \sum_{n=1}^{N}\varphi_{kn}^{t_r,R_x} \tag{4.16}$$

where $\varphi_{kn}^{t_r,R_x} = \tilde{\beta}_{kn}^{t_r,R_x} - \tilde{\beta}_{kn}^{t_r-1}$ is the utility of R_x for U_k to decode the packet X_n and $\mathcal{N}_{R_x}^r$ is the neighboring node set of R_x in the frame t_r.

Step 3 [Relay Selection Scheme Formulation]: Denote the optimally selected relay node set for the transmission frame t_r as $\Omega_r=\left\{R_1^*, R_2^*, \ldots, R_{L_r}^*\right\}$. The relay selection algorithm to select the relay node set for t_r is then formulated as

$$\max_{L_r,\Omega_r} \sum_{i=1}^{L_r} \Phi^{t_r,R_i^*} \tag{4.17}$$

$$s.t.\ \ \mathcal{N}_{R_i^*}^r \cap \mathcal{N}_{R_j^*}^r = \emptyset \quad i \neq j, \quad \forall i, j = 1, 2, \ldots, L_r \tag{4.18}$$

$$\Phi^{t_r,R_i^*} > 0, \quad \forall i = 1, 2, \ldots, L_r. \tag{4.19}$$

Note that at the beginning of the data dissemination, only the original source, i.e., RSU has the dissemination utility. Therefore, the relay node selected in the set Ω_1 for the first data dissemination cycle will definitely be an RSU with the designed relay selection strategy.

Step 4 [Decoding Status Updating]: After the relay selection procedure for the transmission frame t_r, the decoding status at vehicle U_k is predicted and updated as follows:

$$\tilde{\beta}_{kn}^{t_r} = \begin{cases} \tilde{\beta}_{kn}^{t_r, R_k^r}, & if\ U_k \in \mathcal{N}_{R_k^r}^r\ and\ R_k^r \in \Omega_r \\ \tilde{\beta}_{kn}^{t_r - 1}, & otherwhise \end{cases} \tag{4.20}$$

The relay selection for the next transmission frame t_{r+1} is then based on $\tilde{\beta}_{kn}^{t_r}$ with the designed relay selection strategy.

4.4.1.4 Relay Transmission Phase

If a node is selected as a relay for the transmission frame t_r, then it relays the data which have been correctly decoded to its neighbors with STNC. Given that $R_k^r (R_k^r \in \{\mathbf{RSU} \cup \mathbf{U}\})$ is the selected relay node that is adjacent to U_k in the transmission frame t_r, the received signal at U_k can be given as

$$Y_k^r(t) = \sqrt{P_k^r}\sqrt{L_k^r} h_k^r X_C^{rk}(t) + \omega_k^r(t) \tag{4.21}$$

where P_k^r denotes the transmit power at R_k^r, L_k^r and h_k^r denote the large and small scale fading coefficients of the channel between U_k and R_k^r in the transmission frame t_r, respectively, $\omega_k^r(t)$ is the noise at the receiver, and the data $X_C^{rk}(t) = \sum_{n=1}^{N} X_n \alpha_n(t) \beta_{rkn}^{t_{r-1}}$ is the coded packet with the STNC method at R_k^r, where $\beta_{rkn}^{t_{r-1}} = \{0, 1\}$ stands for the decoding state of packet X_n at R_k^r at the end of former transmission frame t_{r-1}. $\beta_{rkn}^{t_{r-1}} = 1$ means successful decoding while $\beta_{rkn}^{t_{r-1}} = 0$ the opposite. If R_k^r is a RSU, we have $\beta_{rkn}^{t_{r-1}} \equiv 1$. Furthermore, we assume the channel coefficient h_k^r is i.i.d. complex Gaussian with mean zero and variance σ_h^2, i.e., $h_k^r \sim \mathcal{CN}(0, \sigma_h^2)$. The noise $\omega_k^r(t)$ is also Gaussian with mean zero and variance N_0.

4.4.1.5 Feedback Phase

In order to eliminate the transmission collision problem in the feedback phase, each vehicle is allocated with a dedicated sub-channel in the control channel band and a feedback frame to update the information. Specifically, assuming that there are B RBs in the control channel, the kth vehicle node U_k then adopts the $b(= mod(k, B) + 1)$th, $b = 1, 2, \ldots, B$ RB and the $e(= \lceil k/B \rceil)$th, $e = 1, 2, \ldots, E$ transmission frame for the feedback information. The duration ξ should be set over $E\ell(= \lceil K/B \rceil \ell)$ to ensure all the vehicles have transmission opportunities in the feedback phase.

Moreover, when a vehicle switches from one CS control range to another CS control range during one dissemination cycle, the new CS then would hear the feedback information from the vehicle while the previous CS would not. Under this condition, the old CS will unregister those vehicles, while the new CS will register them.

4.4.1.6 Signal Detection

The matched-filtering and maximal-ratio combining (MRC) methods [49] are adopted for the signal detection. For the vehicle U_k to decode the packet X_n from the received signal sent by R_k^r within the transmission frame t_r, it first applies a bank of matched-filters to the signals using signature waveforms

$$\begin{aligned} Y_{kn}^r &= \left\langle Y_k^r(t), \alpha_m(t)\right\rangle \\ &= \sqrt{P_k^r}\sqrt{L_k^r} h_k^r \sum_{n=1}^{N} X_n \beta_{rkn}^{t_{r-1}} \rho_{mn} + \omega_{km}^r. \end{aligned} \tag{4.22}$$

Then, it forms an $N \times 1$ vector comprised of the Y_{kn}^r as

$$\mathbf{Y}_k^r = \sqrt{P_k^r}\sqrt{L_k^r} h_k^r \mathbf{R}\mathbf{B}_{rk}^{t_{r-1}}\mathbf{X} + \mathbf{w}_k^r \tag{4.23}$$

where $\mathbf{Y}_k^r = \left[Y_{k1}^r, Y_{k2}^r, \ldots, Y_{kN}^r\right]^{\mathcal{T}}$, $\mathbf{B}_{rk}^{t_{r-1}} = diag\left\{\beta_{rk1}^{t_{r-1}}, \beta_{rk2}^{t_{r-1}}, \ldots, \beta_{rkN}^{t_{r-1}}\right\}$, $\mathbf{w}_k^r = \left[\omega_{k1}^r, \omega_{k2}^r, \ldots, \omega_{kN}^r\right]^{\mathcal{T}} \sim \mathcal{CN}(\mathbf{0}, N_0\mathbf{R})$, and

$$\mathbf{R}=\begin{bmatrix} 1 & \rho_{12} & \cdots & \rho_{1N} \\ \rho_{21} & 1 & \cdots & \rho_{2N} \\ \vdots & \vdots & \ddots & \vdots \\ \rho_{N1} & \rho_{N2} & \cdots & 1 \end{bmatrix}. \tag{4.24}$$

The signal vector $\mathbf{Y}_k^r$ is then decorrelated to obtain

$$\tilde{\mathbf{Y}}_k^r = \mathbf{R}^{-1}\mathbf{Y}_k^r = \sqrt{P_k^r}\sqrt{L_k^r} h_k^r \mathbf{B}_{rk}^{t_{r-1}}\mathbf{X} + \tilde{\mathbf{w}}_k^r \tag{4.25}$$

where $\tilde{\mathbf{w}}_k^r \sim \mathcal{CN}(\mathbf{0}, N_0\mathbf{R}^{-1})$. The detected signal X_n at U_k, i.e., the nth element of the vector $\tilde{\mathbf{Y}}_k^r$ is

$$\tilde{Y}_{kn}^r = \sqrt{P_k^r}\sqrt{L_k^r} h_k^r \beta_{rkn}^{t_{r-1}} X_n + \tilde{\omega}_{kn}^r \tag{4.26}$$

where $\tilde{\omega}_{kn}^r \sim \mathcal{CN}(0, N_0\varepsilon_n)$ with ε_n being the nth diagonal element of the matrix $\mathbf{R}^{-1}$ associated with the packet X_n.

Since U_k can receive r same relaying signals from the transmission frame t_1 to t_r, it can further detect the desired packet X_n by using a MRC detector. The final detected packet X_n is then given by

$$\hat{Y}_{kn}^r \triangleq a_{kn}^r X_n + \hat{\omega}_{kn}^r \tag{4.27}$$

where

$$a_{kn}^{r} = \sum_{i=1}^{r} \frac{P_k^i L_k^i \left\| h_k^i \right\|^2 \beta_{ikn}^{t_{i-1}}}{N_0 \varepsilon_n} \tag{4.28}$$

and $\hat{\omega}_{kn}^{r} \sim \mathcal{CN}(0, a_{kn}^{r})$. Note that when the relay nodes in all the R frames fail to decode X_n, i.e., $\beta_{ikn}^{t_{i-1}} = 0$, U_k then cannot correctly decode X_n.

According to the above descriptions, the overall procedure of the centralized data dissemination scheduling protocol is provided in Fig. 4.20.

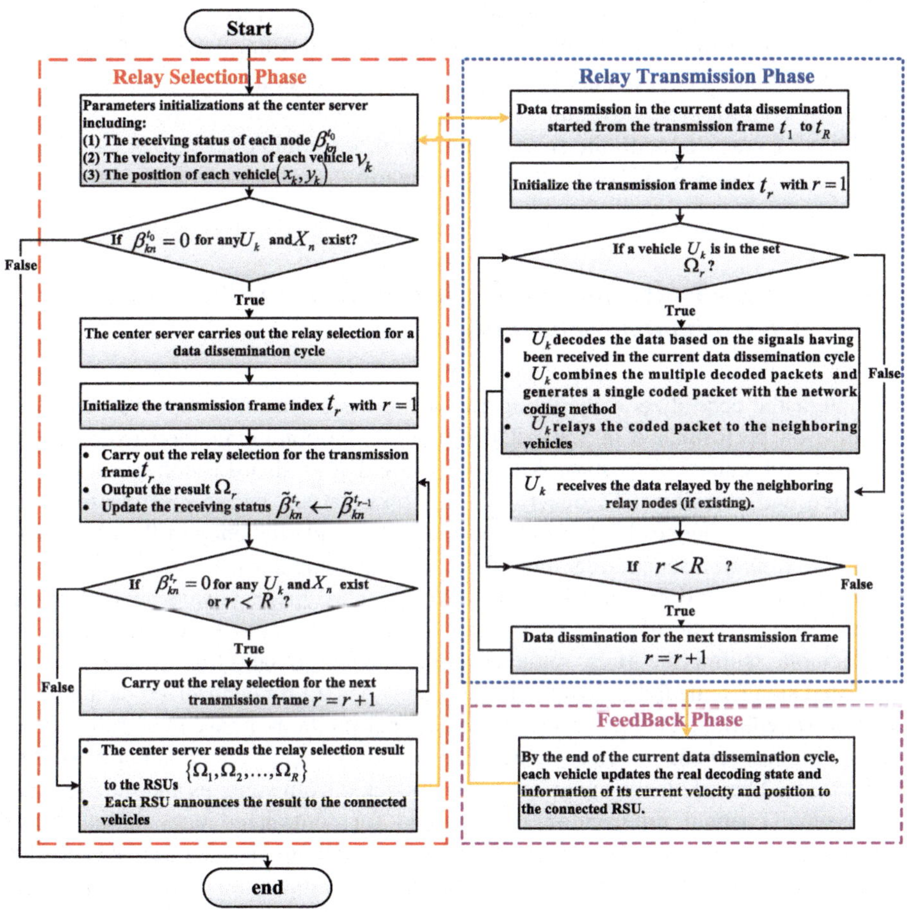

Fig. 4.20 The overall diagram of the proposed data dissemination strategy

4.4.1.7 Performance Evaluation

In [45], the efficiency of the centralized data dissemination scheduling protocol was evaluated in both highway and urban vehicular scenarios. The AoI in the highway scenario consists of a bi-direction, four lane highway with length 6 km, whereas the AoI in the urban scenario is 4×4 km. In Fig. 4.21, the average downloading delay performance comparisons among the centralized data dissemination scheduling approach, the CodeOn-Basic given in [41], and the RSU broadcasting only scheme are provided. From Fig. 4.21, it can be seen that the centralized data dissemination scheduling approach achieves significantly reduced data dissemination delay even in a dense vehicular network, with a proper pre-set ρ, which is defined as the cross-correlation value of the non-orthogonal code in STNC. Moreover, the centralized data dissemination scheduling approach results in even better delay performance in dense vehicular networks than in sparse vehicular networks. This is because under the dense condition, the advantage of the space diversity in the STNC can be further exploited. Compared with the sparse network, more suitable nodes can be selected for data relaying with STNC in the dense scenario.

4.4.2 Large-Scale Channel Prediction-Based Data Dissemination Scheduling

Although the centralized scheduling can achieve significantly improved data dissemination performance in vehicular networks, the communication overhead for CSI and vehicular information is very high, and it is also difficult to collect the accurate and real-time CSI due to the fast time-variant fading and the feedback delay. These challenges lead to reduced feasibility and efficiency of the centralized data dissemination scheduling applied in practical vehicular networks.

Actually, we find that in vehicular networks, the vehicles always have relatively stable moving patterns in a short time, which makes the states and locations predictable within the time scale of communication data frame, and severely outdated small-scale fading would not bring any benefit for centralized data dissemination scheduling. Therefore, in [50], a channel prediction-based centralized data dissemination scheme was proposed, in which only large-scale channel prediction is performed based on current information collected from vehicles and the predicted large-scale channel information is exploited for centralized data dissemination scheduling. The proposed large-scale channel prediction-based centralized scheduling scheme can be realized with both low communication overhead and low complexity, making centralized data dissemination scheduling feasible for vehicular networks. In addition, as illustrated in Fig. 4.22, the large-scale channel prediction-based centralized scheduling can be easily integrated with current centralized data dissemination protocols by introducing the prediction process in the information collection phase.

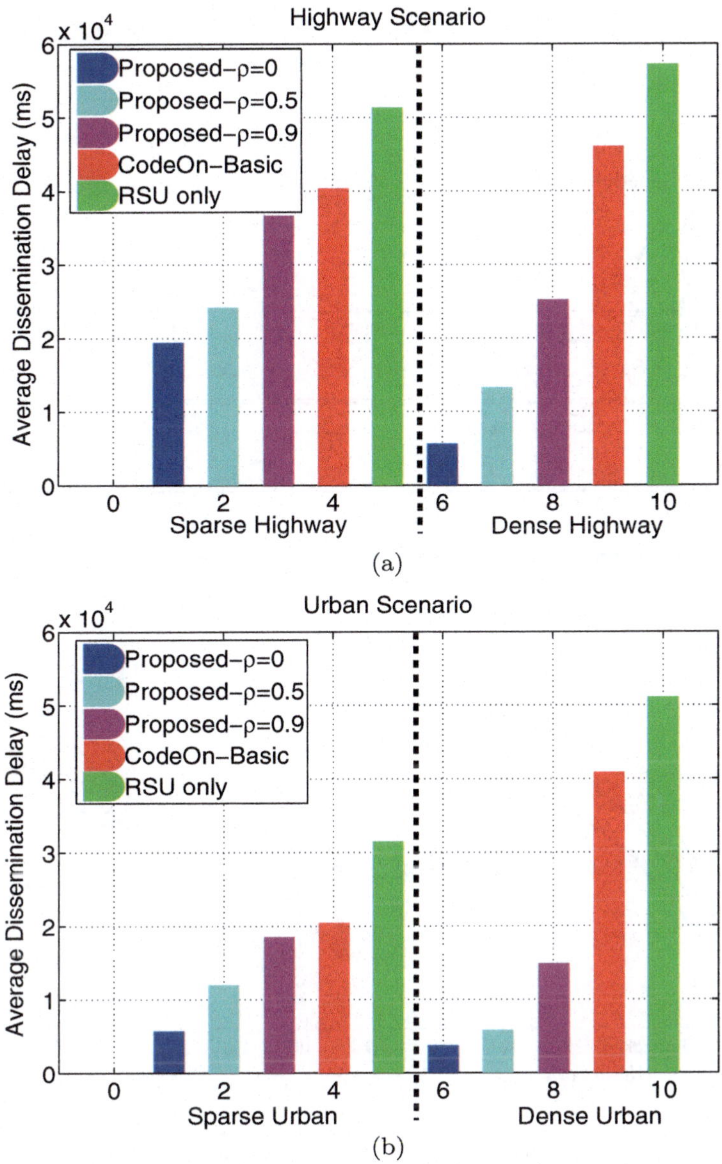

Fig. 4.21 Average downloading delay performance comparison for data dissemination in vehicular networks. (**a**) Highway scenario. (**b**) Urban scenario

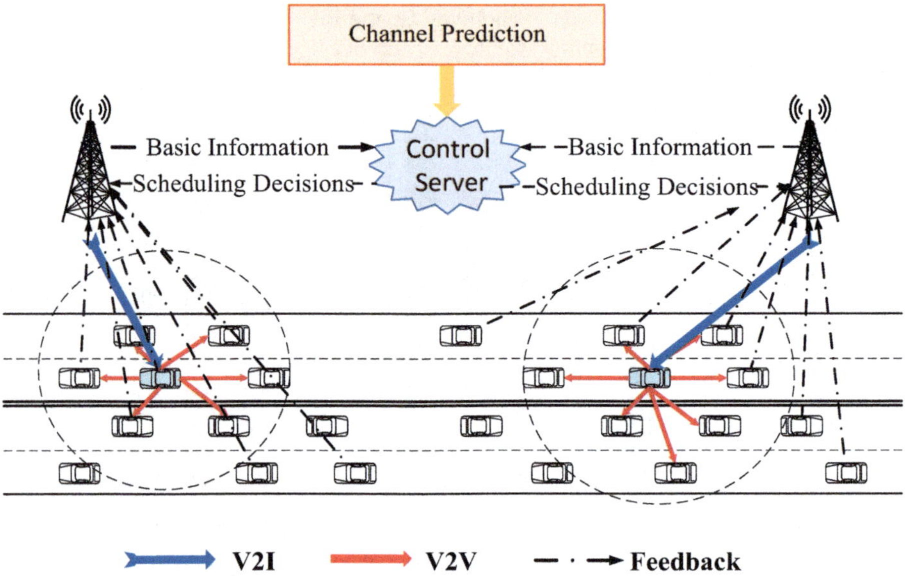

Fig. 4.22 Throughput performance comparisons

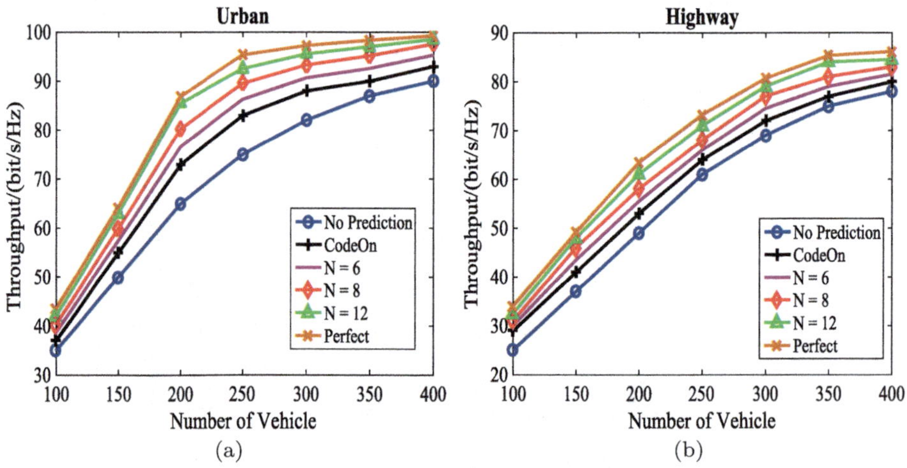

Fig. 4.23 The throughput performance comparison in (**a**) Urban (**b**) Highway

As shown in Fig. 4.23, the large-scale channel prediction-based centralized scheduling scheme can achieve very close performance to the perfect CSI one, where in the perfect CSI scheme, the control server knows the exact and real-time CSI (including both large and small scale fading) for every scheduling frame. This verifies the efficiency of the proposed large-scale channel prediction scheme in realizing centralized data dissemination scheduling in practical vehicular networks.

4.5 The Next Leap

In future 5G-enabled VCN, the network topology will become more complex due to the increasing communications modes (known as vehicle to everything) in the same heterogeneous network, thus leading to more complicated accessing and interference scenario. Moreover, the QoS requirements will be more stringent, e.g., much higher network throughput and lower accessing/transmission latency. Therefore, improved MAC design is required for future VCN.

Non-orthogonal multiple access (NOMA) has been regarded as an essential enabling technology for the 5G wireless systems to meet the heterogeneous demands on low latency, high reliability, massive connectivity, improved fairness, and high throughput. All these align well with the QoS requirements of V2X networks. The key idea behind NOMA is to serve multiple users in the same resource block, such as a time slot, subcarrier, or spreading code [51]. There are two dominant NOMA solutions for future networks, namely the power domain multiplexing and the code domain multiplexing. For power domain multiplexing, different users are allocated with different power levels according to their channel conditions, and the successive interference cancellation (SIC) is used to mitigate multi-user interference. For code domain multiplexing, e.g., sparse code multiple access, different users are assigned with different codes, and then multiplexed over the same time-frequency resources. By exploiting the concept of NOMA together with the above introduced D2D-based resource sharing mechanism, effective and efficient MAC designs with significantly improved network performance can be expected.

Supported by the big data technology, data-driven network resource allocation [52] is also a promising means to effectively achieve improved short-term network optimization and long-term network management.

References

1. Wireless LAN Medium Access Control (MAC) and Physical Layer (PHY) Specifications Amendment 6: Wireless Access in Vehicular Environments, IEEE Standard 802.11p, 2010.
2. IEEE Standard for Information technology–Local and metropolitan area networks–Specific requirements–Part 11: Wireless LAN Medium Access Control (MAC) and Physical Layer (PHY) Specifications - Amendment 8: Medium Access Control (MAC) Quality of Service Enhancements," in IEEE Std 802.11e-2005, pp.1–212, Nov. 2005
3. X. Shen, X. Cheng, R. Zhang, B. Jiao, and Y. Yang, "Distributed Congestion Control Approaches for the IEEE 802.11p Vehicular Networks," *IEEE Intelligent Transportation Systems Magazine*, vol. 5, no. 4, pp. 50–61, winter 2013.
4. X. Shen, R. Zhang, X. Cheng, Y. Yang, and B. Jiao, "Distributed multi-priority congestion control approach for IEEE 802.11p vehicular networks," in *Proc. 2012 12th International Conference on ITS Telecommunications*, Taipei, 2012, pp. 93–97.
5. L. Wischhof and H. Rohling, "Congestion control in vehicular ad hoc networks," in *Proc. IEEE ICVES 2005*, Xian, China, Oct. 2005.

6. C. Huang, Y. P. Fallah, R. Sengupta, and H. Krishnan, "Information dissemination control for cooperative active safety applications in vehicular ad-hoc networks," in *Proc. IEEE GLOBECOM 2009*, Honolulu, Hawaii, USA, Nov. 30 2009–Dec. 4 2009.
7. F. Ye, R. Yim, S. Roy, and J. Zhang, "Efficiency and reliability of one-hop broadcasting in vehicular ad hoc networks," *IEEE J. Sel. Areas Commun.*, vol. 29, no. 1, pp. 151–160, Jan. 2011.
8. C. Chuang and S. Kao, "A probabilistic discard congestion control for safety information in vehicle-to-infrastructure vehicular network," in *Proc. 40th International Conference on CIE 2010*, Awaji City, Japan, Jul. 2010.
9. J. He, H. Chen, T. M. Chen, and W. Cheng, "Adaptive congestion control for DSRC vehicle networks," *IEEE Commun. Lett.*, vol. 14, no. 2, pp. 127–129, Feb. 2010.
10. C. Hsu, C. Hsu and H. Tseng, "MAC channel congestion control mechanism in IEEE 802.11p/WAVE vehicle networks," in *Proc. IEEE VTC 2011-Fall*, San Francisco, USA, Sept. 2011.
11. H. Jang and W. Feng, "Network status detection-based dynamic adaptation of contention window in IEEE 802.11p," in *Proc. IEEE VTC 2010-Spring*, Taipei, Taiwan, May. 2010.
12. Y. Zang, L. Stibor, X. Cheng, H. J. Reumerman, A. Paruzel, and A. Barroso, "Congestion control in wireless networks for vehicular safety applications", in *Proc. The 8th European Wireless Conference*, Paris, France, Apr. 2007.
13. M. Barradi, A. S. Hafid, and J. R. Gallardo, "Establishing strict priorities in IEEE 802.11p WAVE vehicular networks," in *Proc. IEEE GLOBECOM 2010*, Miami, Florida, USA, Dec. 2010.
14. M. S. Bouassida and M. Shawky, "A cooperative and fully-distributed congestion control approach within VANETs," in *Proc. IEEE ITSC 2009*, St. Louis, Missouri, USA, Oct. 2009.
15. R. Stanica, E. Chaput, and A. Beylot, "Congestion control in CSMA-based vehicular networks: Do not forget the carrier sensing," in *Proc. IEEE SECON 2012*, Seoul, Korea, June. 2012.
16. M. Torrent-Moreno, J. Mittag, P. Santi, and H. Hartenstein, "Vehicle-to-vehiclecommunication: fair transmit power control for safety-critical information", *IEEE Trans. Veh. Technol.*, vol. 58, pp. 3684–3703, Sep. 2009.
17. Y. P. Fallah, C. Huang, R. Sengupta, and H. Krishnan, "Congestion control based on channel occupancy in vehicular broadcast networks," in *Proc. IEEE VTC 2010-Fall*, Ottawa, ON, Canada, Sept. 2010.
18. L. Wei, X. Xiao, Y. Chen, M. Xu, and H. Fan, "Power-control-based broadcast scheme for emergency messages in VANETs," in *Proc. ISCIT 2011*, Hangzhou, China, Oct. 2011.
19. O. Chakroun, S. Cherkaoui, and J. Rezgui, "MUDDS: multi-metric unicast data dissemination scheme for 802.11p VANETs," in *Proc. IWCMC 2012*, Limassol, CYPRUS, Aug. 2012.
20. C. Huang, Y. P. Fallah, R. Sengupta, and H. Krishnan, "Adaptive intervehicle communication control for cooperative safety systems," *IEEE Network*, vol. 24, no. 1, pp.6–13, Jan.–Feb. 2010.
21. W. Guan, J. He, L. Bai, and Z. Tang, "Adaptive congestion control of DSRC vehicle networks for collaborative road safety applications," in *Proc. IEEE LCN 2011*, Sydney, Australia, Oct. 2011.
22. M. Sepulcre, J. Gozalvez, J. Härri, and H. Hartenstein, "Contextual communications congestion control for cooperative vehicular networks," *IEEE Trans. Wireless Commun.*, vol. 10, no. 2, pp. 385–389, Feb. 2011.
23. S. Djahel and Y. Ghamri-Doudane, "A robust congestion control scheme for fast and reliable dissemination of safety messages in VANETs," in *Proc. IEEE WCNC 2012*, Paris, France, Apr. 2012.
24. R. Cao and L. Yang, "The affecting factors in resource optimization for cooperative communications: A case study," *IEEE Trans. Wireless Commun.*, vol. 11, no. 12, pp. 4351–4361, Dec. 2012.
25. M. Sepulcre and J. Gozalvez, "Wireless vehicular adaptive radio resource management policies in congested channels," in *Proc. ISWCS 2007*, Trondheim, Norway, Oct. 2007.

26. M. Torrent-Moreno, P. Santi, and H. Hartenstein, "Fair sharing of bandwidth in VANET," in *Proc. 2nd ACM Int. Workshop VANET*, Cologne, Germany, Sept. 2005.
27. X. Cheng, L. Yang, and X. Shen, "D2D for Intelligent Transportation Systems: A Feasibility Study," *IEEE Transactions on Intelligent Transportation Systems*, vol. 16, no. 4, pp. 1784–1793, Aug. 2015.
28. R. Zhang, X. Cheng, L. Yang, X. Shen, and B. Jiao, "A novel centralized TDMA-based scheduling protocol for vehicular networks," *IEEE Transactions on Intelligent Transportation Systems*, vol. 16, no. 1, pp. 411–416, Feb. 2015.
29. R. Zhang, X. Cheng, Q. Yao, C.-X. Wang, Y. Yang, and B. Jiao, "Interference graph-based resource-sharing schemes for vehicular networks," *IEEE Transactions on Vehicular Technology*, vol. 62, no. 8, pp. 4028–4039, Oct. 2013.
30. K. Doppler, M. Rinne, C. Wijting, C. Ribeiro, and K. Hugl, "Device-to-device communication as an underlay to LTE-Advanced networks," *IEEE Commun. Mag.*, vol. 47, no. 12, pp. 42–49, Dec. 2009.
31. R. Zhang, X. Cheng, L. Yang and B. Jiao, "Interference Graph-Based Resource Allocation (InGRA) for D2D Communications Underlaying Cellular Networks," *IEEE Transactions on Vehicular Technology*, vol. 64, no. 8, pp. 3844–3850, Aug. 2015.
32. H. Min, W. Seo, J. Lee, S. Park, and D. Hong, "Reliability Improvement Using Receive Mode Selection in the Device-to-Device Uplink Period Underlaying Cellular Networks," *IEEE Transactions on Wireless Communications*, vol. 10, no. 2, pp. 413–418, Feb. 2011.
33. C. H. Yu, K. Doppler, C. B. Ribeiro and O. Tirkkonen, "Resource Sharing Optimization for Device-to-Device Communication Underlaying Cellular Networks," *IEEE Transactions on Wireless Communications*, vol. 10, no. 8, pp. 2752–2763, Aug. 2011.
34. L. Fang, R. Zhang, X. Cheng, J. Xiao, and L. Yang, "Cooperative Content Download-and-Share: Motivating D2D in Cellular Networks," *IEEE Communications Letters*, vol. 21, no. 8, pp. 1831–1834, Aug. 2017.
35. T. Yang, R. Zhang, X. Cheng, and L. Yang, "Graph coloring based resource sharing (GCRS) scheme for D2D communications underlaying full-duplex cellular networks," *IEEE Transactions on Vehicular Technology*, vol. 66, no. 8, pp. 7506–7517, Aug. 2017.
36. Y. Zhu, R. Zhang, X. Cheng, and L. Yang, "An interference-free graph based TDMA scheduling protocol for vehicular ad-hoc networks," in *Proc. IEEE 85th Vehicular Technology Conference (VTC Spring)*, Sydney, NSW, 2017.
37. A. Jalali, R. Padovani, and R. Pankaj, "Data throughput of CDMA-HDR a high efficiency-high data rate personal communication wireless system," in *Proc. VTC 2000-Spring*, Tokyo, Japan, May 2000.
38. E. Karamad and F. Ashtiani, "A modified 802.11-based MAC scheme to assure fair access for vehicle-to-roadside communications," *Computer Communications*, vol. 31, no. 12, pp. 2898–2906, Jul. 2008.
39. G. Bianchi, "Performance analysis of the IEEE 802.11 distributed coordination function," *IEEE Journal on Selected Areas in Communications*, vol. 18, no. 3, pp. 535–547, Mar. 2000.
40. U. Lee, J. S. Park, J. Yeh, G. Pau, and M. Gerla, "Code torrent: Content distribution using network coding in VANET," in *Proc. MobiShare*, Los Angeles, CA, USA, Sept. 2006.
41. M. Li, Z. Yang, and W. Lou, "CodeOn: Cooperative popular content distribution for vehicular networks using symbol level network coding," *IEEE J. Sel. Areas Commun.*, vol. 29, no. 1, pp. 223–235, Jan. 2011.
42. M. Sardari, F. Hendessi, and F. Fekri, "DMRC: Dissemination of multimedia in vehicular networks using rateless codes," in *Proc. IEEE INFOCOM Workshops*, Rio De Janeiro, Brazil, Apr. 2009, pp. 19–25.
43. C. Stefanovic, D. Vukobratovic, F. Chiti, L. Niccolai, V. Crnojevic, and R. Fantacci, "Urban infrastructure-to-vehicle traffic data dissemination using UEP rateless codes," *IEEE J. Sel. Areas Commun.*, vol. 29, no. 1, pp. 94–102, Jan. 2011.
44. V. Palma, E. Mammi, A. M. Vegni, and A. Neri, "A fountain codes-based data dissemination technique in vehicular ad-hoc networks," in *Proc. ITST 2011*, St. Petersburg, Russia, Aug. 2011, pp. 750–755.

45. X. Shen, X. Cheng, L. Yang, R. Zhang, and B. Jiao, "Data dissemination in VANETs: A scheduling approach," *IEEE Transactions on Intelligent Transportation Systems*, vol. 15, no. 5, pp. 2213–2223, Oct. 2014.
46. H. Q. Lai and K. J. R. Liu, "Spacetime network coding," *IEEE Trans. Signal Process.*, vol. 59, no. 4, pp. 1706–1718, Apr. 2011.
47. C. Fragouli, J. Widmer, and J. Y. Le Boudec, "Efficient broadcasting using network coding," *IEEE/ACM Trans. Netw*, vol. 16, no. 2, pp. 450–463, Apr. 2008.
48. N. Dong, T. Tran, N. Thinh, and B. Bose, "Wireless broadcast using network coding," *IEEE Trans. Veh. Technol.*, vol. 58, no. 2, pp. 914–925, Feb. 2009.
49. H. Q. Lai and K. J. R. Liu, "Space-time network coding," *IEEE Trans. Signal Process.*, vol. 59, no. 4, pp. 1706–1718, Apr. 2011.
50. F. Zeng, R. Zhang, X. Cheng, and L. Yang, "Channel prediction based scheduling for data dissemination in VANETs," *IEEE Communications Letters*, vol. 21, no. 6, pp. 1409–1412, Jun. 2017.
51. Z. Ding, X. Lei, G. K. Karagiannidis, R. Schober, J. Yuan, and V. K. Bhargava, "A Survey on Non-Orthogonal Multiple Access for 5G Networks: Research Challenges and Future Trends," *IEEE Journal on Selected Areas in Communications*, vol. 35, no. 10, pp. 2181–2195, Oct. 2017.
52. Y. Bao, H. Wu, and X. Liu, "From Prediction to Action: Improving User Experience With Data-Driven Resource Allocation," *IEEE Journal on Selected Areas in Communications*, vol. 35, no. 5, pp. 1062–1075, May 2017.

Chapter 5
Wireless-Vehicle Integration: VCN-Based Applications

While the wireless-vehicle combination intends to increase the efficiency of link-level and network-level data transmissions to fulfill the communication requirements in vehicular applications, the wireless-vehicle integration focuses on exploring the core functions of vehicles that are evolving more and more towards being highly intelligent and electrified. Surrounding the core vehicle functions, in this chapter, the requirements on the supporting wireless infrastructure and how to achieve these requirements will be discussed from the wireless-vehicle integration perspective. Specifically, we will focus on some interesting VCN-based vehicular applications including electric vehicles, distributed data storage, and physical layer security. As for the next leap, VCN-based autonomous driving is also discussed.

5.1 VCN-Based Applications

As vehicles become increasingly intelligent and energy-aware, the vehicle functions are growing increasingly dependent on effective and efficient communications. For instance, when the traditional automatic driving assistance systems (ADAS) evolve towards autonomous driving, it is expected that enormous amount of data exchange among vehicles and control centers would emerge in order to fulfill driving environment acquisition, vehicle coordination, as well as transportation planning and optimization. Another example involves electric vehicles (EVs), for which the bi-directional charging and discharging need to be scheduled together with route planning and traffic control. In these scenarios, simply combining communications with vehicles would not satisfy the increasingly stringent requirements of the vehicle functions. Instead, wireless and vehicles need to be intimately integrated. In other words, the wireless system design has to be carefully tailored for core vehicle functions, whereas the design of vehicle functions has to account for the wireless limitations as well. Figure 5.1 illustrates some important VCN-based applications.

X. Cheng et al., *5G-Enabled Vehicular Communications and Networking*, Wireless Networks, https://doi.org/10.1007/978-3-030-02176-4_5

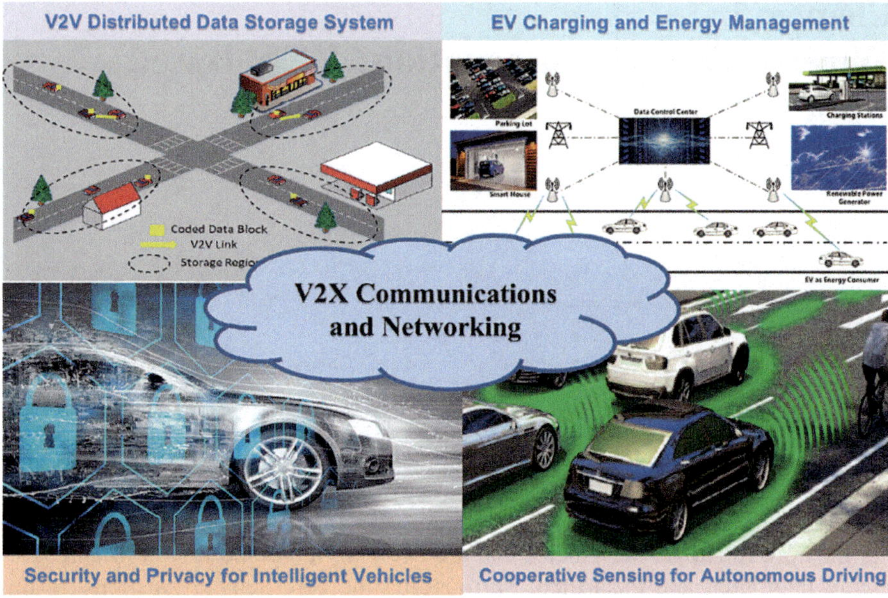

Fig. 5.1 Wireless-vehicle integration for various vehicular applications

5.2 Electric Vehicles (EVs)

With ever increasing concerns on environmental issues and clean energy, EVs are regarded as one of the most effective strategies to reduce fuel dependence and gas emissions and to increase the efficiency of energy conversion, and have attracted more and more attention from government, industry, and consumers [1]. In this section, based on the real-time and efficient information exchange facilitated by 5G-enabled VCN, we discuss an effective energy management framework for EVs and the smart grid, and then provide a cooperative V2V charging protocol.

5.2.1 EV-Integrated Vehicular Networks

When integrated into the smart grid via the charging and discharging operations, EVs can not only serve as a transportation tool but also act as controllable loads and distributed energy sources [2]. Specifically, EVs can exchange energy with the power grid, smart houses, and other EVs through bidirectional vehicle-to-grid (V2G), vehicle-to-home (V2H), and vehicle-to-vehicle (V2V) technologies [3], respectively, making EVs become both energy consumers and providers in vehicular networks. The efficiency of charging and routing decisions of EVs is determined by the real-time and efficient coordination and scheduling. This requires enhanced vehicular communications and networking with low latency and high reliability. On the other hand, the unique behaviors of EVs may also affect the network topology and thus the communications design therein, e.g., gathering of EVs at a specific spot (charging stations or parking lots) for quite a while (since the charging/discharging process cannot be completed in a short time), which may bring joint optimization to improve V2X communications and data sharing among vehicles.

In recent years, as a promising concept for future energy system, Internet of Energy (IoE) is described as an Internet of Things (IoT) woven together into a choreographed dance of energy by a humming buzz of power lines and wireless communications [4]. In IoE, renewable-energy power plants, transmission links, electrical meters, appliances, and the moving EVs will be able to talk to each other in real time about the electrical loads and energy prices and share power with each other if demanded [5]. Just like the information Internet forever changed the way information is made, shared, and stored, IoE will also change the way we produce, distribute, and store energy. IoE is envisioned as a smart architecture that enables flexible energy sharing among the involved units. The mobile and energy storage features make the EVs play an important role in increasing the flexibility and possibilities of power transfer in the IoE. On the one hand, the fast development of EVs brings a significant new load on the current power system [6]. Without efficient control strategies, the EV charging process may overload the power grid at peak hours, especially in residential communities. On the other hand, EVs can benefit the power grid as a flexible load through smart charging/discharging scheduling to reduce the peak load and shape the load profile. Therefore, in the literature, with the concept of demand side management (DSM) [7], many works [8–11] have focused on the charging/discharging scheduling and energy management protocols to control and optimize the charging process for EVs integrated with the power grid.

5.2.2 Schedule-Upon-Request Energy Management Framework

Due to the complex and flexible interactions among the EVs and the smart grid, the energy management problem in the power system integrated with EVs is both interesting and challenging. It requires efficient and intelligent coordination of the entire power system to achieve effective energy management. Therefore, in [2, 12], a schedule-upon-request energy management framework was proposed, which facilitates the coordination and schedule of the power transfer among the EVs and the smart grid in a cooperative and intelligent manner. Integrating the power system in the physical space, communications and computations in the cyberspace, and consumer interactions in the social and human dimension, the proposed schedule-upon-request energy management framework is constructed based on a cyber-physical-social (CPS) power system as illustrated in Fig. 5.2.

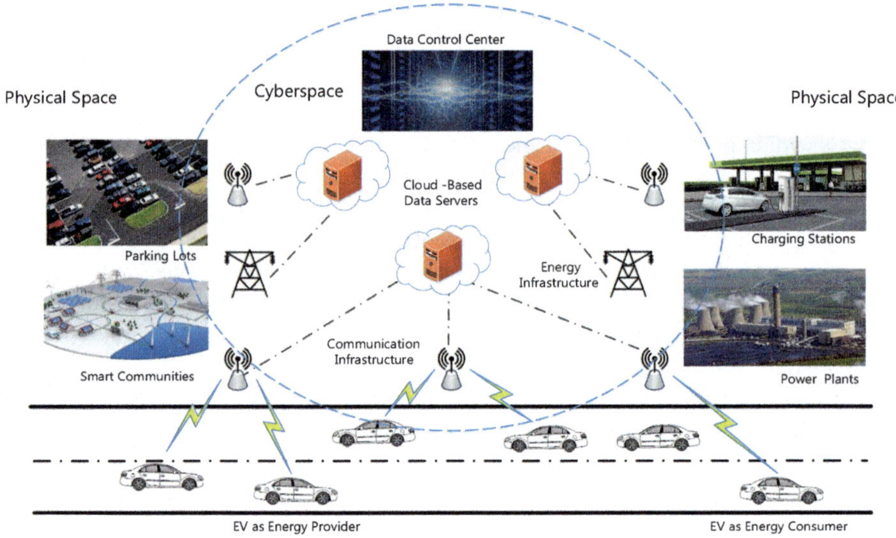

Fig. 5.2 An illustrated EV-integrated CPS power system

The data control center, distributed cloud-based data servers, and the wired and wireless heterogeneous communication networks constitute the backbone of the cyberspace in the schedule-upon-request energy management framework. The data control center handles dynamic and real-time data collection from the physical energy entities under the support of the IoT and intelligent transportation systems (ITS) [13]. The collected data are stored and processed in the cloud-based data servers in a distributed manner. The collected data mainly include the energy surplus or deficit status of each energy entity, the real-time system energy distribution, the grid electricity price, and the individual information of current energy consumers and providers in the system (such as the speed and trip information of the EVs and the locations of the charging stations as well as the community parking lots). On the basis of the collected and stored data, intelligent computations and decisions are carried out to achieve different system coordination and optimization targets. It is necessary to take the mobility information of the EVs for the detailed energy management optimization method design, due to its significant effect on the available power status as well as the additional energy cost to achieve a determined energy trading of the EVs. The energy management decisions will then be fed back to the involved energy entities as control commands for their further operations.

The EVs are both rational energy entities with individual utilities and preferences and they have the incentives to maximize their own utilities through energy trading and power transfer in such an architecture. The utilities of energy providers and consumers that participate in the energy trading and power transfer can be respectively given as

$$\mathcal{U}^{\mathrm{EP}} = \mathcal{R}^{\mathrm{EP}} - \mathcal{C}^{\mathrm{EP}} \tag{5.1}$$

and

$$\mathcal{U}^{\mathrm{EC}} = -\mathcal{C}^{\mathrm{EC}} \tag{5.2}$$

where $\mathcal{R}^{\mathrm{EP}}$ represents the revenue functions of energy providers, and $\mathcal{C}^{\mathrm{EP}}$ and $\mathcal{C}^{\mathrm{EC}}$ denote the cost functions of energy providers and energy consumers, respectively. $\mathcal{R}^{\mathrm{EP}}$ is mostly determined by the trading electricity price and the transferred power amount, which indicate how much revenue the energy entities can collect as energy providers via the energy trading and power transfer with other energy entities as energy consumers. For energy providers, $\mathcal{C}^{\mathrm{EP}}$ essentially includes the additional power consumption as well as the power loss induced by power transfer operations. As for the EVs, this mainly includes the energy cost to drive to the determined energy trading locations. The optimization objective of the energy entities as energy providers is to maximize their profits through active energy trading and power transfer. For energy consumers, $\mathcal{C}^{\mathrm{EC}}$ contains not only the additional power consumption to realize the power transfer operations but also the payment for the received power from other energy entities as energy providers. The optimization objective of the energy entities as energy providers is to minimize their cost in acquiring the demanded power.

As shown in Fig. 5.3, in the schedule-upon-request energy management framework, the energy generation, distribution, and transfer are all under the control of the data control center. The energy entities in the power system update their real-time individual information to the distributed cloud-based data servers. Note that the consumer behaviors and their associated social interactions in the social and human dimension could also be taken into consideration for energy management in the framework. Therefore, the distributed cloud-based data servers will also record the historical load and supply profiles as well as the preferences of each energy entity. During the energy management procedure, the energy consumers (such as EVs and smart communities) that demand power will first send energy requests to the data control center with their embedded communication modules. Based on the received energy requests and the collected and stored information in the cloud-based data servers, the data control center will make intelligent energy management decisions with efficient optimization methods and schedule the power transfer processes accordingly.

5.2.3 Cooperative V2V Charging

Most recently, some works [14–17] have proposed to investigate vehicle-to-vehicle charging strategies, which can offer more flexible charging plans for the EVs in order to offload the EV charging loads from the electric power systems. In the

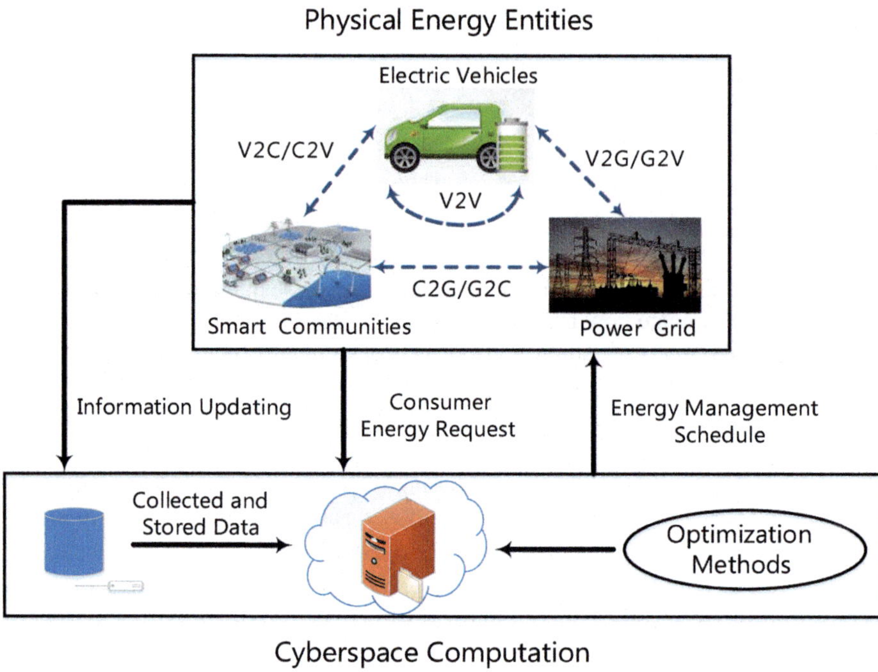

Fig. 5.3 Schedule-upon-request energy management framework

following, we first introduce a developed concept of cooperative V2V charging and then provide a cooperative V2V charging protocol based on V2V matching algorithms.

5.2.3.1 Cooperative V2V Charging Concept

In [15–17], a developed concept based on the V2V operation of V2X concept [3], termed as cooperative V2V charging, was proposed, which describes the power flow connection among different EVs in a cooperative charging/discharging manner. Cooperative V2V charging can enable direct EV-to-EV power transfer through active cooperation among EVs at the energy level. Based on cooperative V2V charging, the charging/discharging behaviors of EVs can be performed in a more flexible and smarter manner.

Cooperative V2V charging is beneficial to both EVs as energy consumers and EVs as energy providers, leading to a win-win energy trading situation. As for EVs as energy consumers, currently, most EVs get charged at the charging stations or after going back home. If a moving EV demands power before it can arrive at the destination, it will have to drive to a nearby charging station to get charged first. However, the current deployment of charging stations is still far from sufficient.

Moreover, the nearby charging station may be in a different direction deviating from the EV's original route to the destination. This causes inconvenience and extra energy consumption for EVs as energy consumers. As an additional feasible charging option, cooperative V2V charging can make the charging behaviors of EVs as energy consumers more flexible and smarter, and thus reduce the drivers' anxiety since their EVs can get charged more easily. As for EVs as energy providers, the EV drivers can make profits through cooperative V2V charging based energy trading with their spare time and surplus power. Especially, with the increasing penetration of renewable energy resources (RESs) such as solar panels in residential houses, many households would have surplus power generated by RESs at a low cost, leading to considerable incentives and motivations on individual energy trading. Even taking the potential cost (e.g., the battery lifetime loss) into consideration, there is still a profit margin for EVs as energy providers to achieve cooperative V2V charging based energy trading with their stored low-cost surplus power. Besides, cooperative V2V charging can also offload the heavy load of the power grid due to the dramatically increasing penetration of EVs in daily life.

Currently, a feasible way to realize V2V power transfer among different EVs is through the V2V framework described in [3], where an aggregator is employed for coordinated control of grouping EVs for charging and discharging. The aggregator behaves as a control device that collects all the information about the EVs and the grid status and then executes the V2V power transfer. Since these aggregators do not need to pull in power from the power grid to operate the V2V power transfer, they would be much cheaper and more easily deployed than the charging stations. For instance, such aggregators can be widely deployed in various communities or public parking lots. One can also envision that in the future IoE, the power transfer among EVs may be achieved via a single charging cable connecting EVs directly or even in a wireless and mobile manner (i.e., wireless V2V power transfer). This will make the charging/discharging among EVs more easily and conveniently. Then, with cooperative V2V charging, EVs will be able to get charged anytime anywhere in the future.

5.2.3.2 EV Utility Definition

First, the utilities of the EVs as energy consumers and energy providers are defined based on their cost/profits through potential energy trading and the corresponding energy cost for driving to the selected trading spot.

5.2.3.3 EV as an Energy Consumer

The utility of EV_i^C, $i \in \mathcal{N}$ as an energy consumer is defined as

$$U_i^C\left(\mathrm{EV}_j^P\right) = -p_t a_i^C - \mathrm{Cost}\left(\mathrm{EV}_i^C, \mathrm{EV}_j^P\right) \tag{5.3}$$

where p_t is the unit power trading price, a_i^C represents the requested power amount, and EV_j^P is the potential paired energy provider for EV_i^C. In general, the electricity buying price p_b set by the power grid for the EVs to trade their surplus power is often considerably lower than the electricity selling price p_s for the EVs to get charged [14]. Based on this, the unit power trading price p_t can be set between the electricity buying and selling prices of the power grid. Therefore, EVs as energy providers can sell their surplus power at a higher price, and EVs as energy consumers can also buy their requested power at a lower price, compared with energy trading through discharging and charging directly with the power grid. In practical applications, the unit power trading price p_t may also vary based on the current information collected at the data control center.

As a preference baseline, the utility of EV_i^C when getting charged at the charging stations is also defined as

$$U_i^C\,(\mathrm{CS}) = -p_s a_i^C - \mathrm{Cost}\left(\mathrm{EV}_i^C, \mathrm{CS}\right) \tag{5.4}$$

where CS denotes the nearest charging station for EV_i^C and p_s is the electricity selling price set by the power grid, that is, the power trading price between the charging station and the EVs as energy consumers.

Note that $\mathrm{Cost}\left(\mathrm{EV}_i^C, \mathrm{EV}_j^P\right)$ and $\mathrm{Cost}\left(\mathrm{EV}_i^C, \mathrm{CS}\right)$ denote the energy cost for EV_i^C to drive to the selected parking lot to achieve power transfer with EV_j^P and to drive to the nearest charging station to get charged, respectively, which can be given as

$$\mathrm{Cost}\left(\mathrm{EV}_i^C, \mathrm{EV}_j^P\right) = p_t \times \beta_i^C \times \mathrm{Dis}\left(\mathrm{EV}_i^C, \mathrm{PL}\right) \tag{5.5}$$

and

$$\mathrm{Cost}\left(\mathrm{EV}_i^C, \mathrm{CS}\right) = p_s \times \beta_i^C \times \mathrm{Dis}\left(\mathrm{EV}_i^C, \mathrm{CS}\right) \tag{5.6}$$

where β_i^C is the moving energy cost per km for EV_i^C, $\mathrm{Dis}\,(x, y)$ represents the driving distance between x and y, and PL denotes the selected parking lot for EV_i^C to achieve power transfer with EV_j^P. Note that here the energy cost for EV_i^C to get charged at the charging station is valued by the electricity selling price p_s of the power grid.

5.2.3.4 EV as an Energy Provider

The utility of EV_j^P, $j \in \mathcal{K}$ as an energy provider is defined as

$$U_j^P\left(\mathrm{EV}_i^C\right) = p_t a_i^C - p_O a_i^C/\eta - \mathrm{Cost}\left(\mathrm{EV}_j^P, \mathrm{EV}_i^C\right)$$

$$- \text{Time}\left(\text{EV}_j^P, \text{EV}_i^C\right) - \Phi\left(\text{EV}_j^P, \text{EV}_i^C\right) \tag{5.7}$$

where p_t and p_0 are the current trading price and the original cost price per unit power, respectively, η represents the V2V power transfer efficiency, $\Phi\left(\text{EV}_j^P, \text{EV}_i^C\right)$ represents the amortized cost to value the battery degradation per each V2V power transfer. $\text{Cost}\left(\text{EV}_j^P, \text{EV}_i^C\right)$ and $\text{Time}\left(\text{EV}_j^P, \text{EV}_i^C\right)$ denote the energy cost and the time cost for EV_j^P to drive to the selected parking lot to achieve power transfer with EV_i^C, respectively, which can be given as

$$\text{Cost}\left(\text{EV}_j^P, \text{EV}_i^C\right) = p_t \times \beta_j^P \times \text{Dis}\left(\text{EV}_j^P, \text{PL}\right) \tag{5.8}$$

and

$$\text{Time}\left(\text{EV}_j^P, \text{EV}_i^C\right) = \theta_j^P \left(\frac{\text{Dis}\left(\text{EV}_j^P, \text{PL}\right)}{v_j^P} + \tau a_i^C/\eta\right) \tag{5.9}$$

where β_j^P is the energy cost per km for EV_j^P, θ_j^P represents the value of time for EV_j^P, v_j^P is the velocity of EV_j^P, and τ denotes the V2V power transfer speed per unit of power. Here we assume that the current surplus power amount of EV_j^P for energy trading denoted by a_j^P satisfies $a_j^P \geq a_i^C$.

As for the time cost, one should note that how to value such a time cost objectively is difficult in practical applications, since the value of time for different people would be quite different and highly subjective. Similar to Uber drivers, the EV drivers who send their information to the control center to act as energy providers may mostly be the ones with some spare time and willing to make profits with their surplus power via energy trading. Therefore, in our opinion, the effect of the time cost should be much smaller than that of the energy cost. As for the amortized cost due to battery degradation, the value of Φ in practical applications can be calculated based on the information reported by the EVs and stored in the data control center. Referring to [18], the battery degradation cost model takes both the battery replacement cost and the battery wear-out via energy processing into consideration, and then $\Phi\left(\text{EV}_j^P, \text{EV}_i^C\right)$ can be given as

$$\Phi\left(\text{EV}_j^P, \text{EV}_i^C\right) = \phi_j \times D_j \times a_i^C \tag{5.10}$$

where ϕ_j represents the battery replacement cost of EV_j^P, and D_j is the capacity degradation coefficient.

Since the electricity buying price p_b set by the power grid for EVs as energy providers to trade their surplus power is lower than the power trading price p_t via cooperative V2V charging, EVs as energy providers would prefer to trade with EVs as energy consumers instead of the power grid, if a positive utility can be achieved.

5.2.3.5 Cooperative V2V Charging Protocol Based on V2V Matching Algorithms

Based on the defined utilities of EVs as energy consumers and energy providers, we provide a flexible energy management protocol for the EVs in this section. The flow chart of the procedure of the proposed energy management protocol is given in Fig. 5.4. The data control center acts as the central controller for the energy management process. During the procedure, the data control center collects and updates the real-time information via mobile Internet periodically. The collected information at the data control center includes the real-time location and moving information from EVs, the location information of the nearby charging stations, smart houses, and parking lots, the charging request and demanded power amount from EVs as energy consumers, the available trading power amount from EVs as energy providers, and the real-time unit electricity price from the charging stations. Based on the collected information, the data control center performs a selected

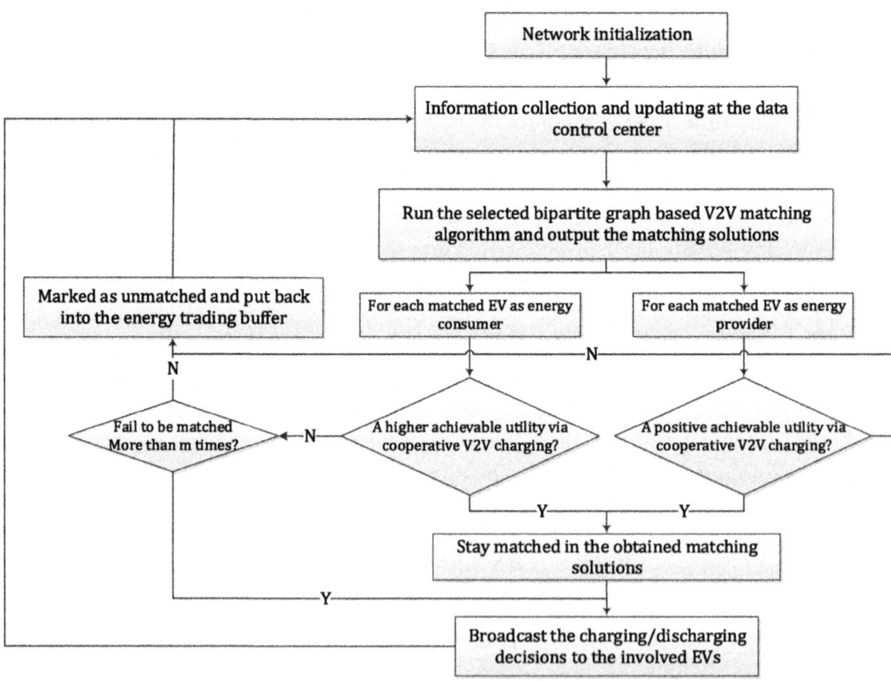

Fig. 5.4 Flow chart of the cooperative V2V charging protocol procedure at the data control center

bipartite graph based V2V matching algorithm to obtain an efficient and effective V2V matching and help the EVs make smart charging/discharging decisions.

In order to find the optimal EV pairs for cooperative V2V charging, three effective and efficient V2V matching algorithms were further provided, that is, max-weight V2V matching that can achieve optimized network social welfare (i.e., the sum edge weight of all the matched EV pairs), EV-consumer-oriented stable V2V matching, and EV-provider-oriented stable V2V matching. Note that although the provided EV-consumer-oriented and EV-provider-oriented V2V matching algorithms can both realize stable matchings, they have significant consequences. The EV-consumer-oriented algorithm yields an EV-consumer-optimal stable matching, in which each EV as energy consumer has the best matched partner that it can have in any stable matching, whereas the EV-provider-oriented algorithm leads to an EV-consumer-optimal output. This property is referred to as the polarization of stable matchings [19].

During the V2V matching process, the data control center will automatically choose a best available parking lot for each potential paired EV based on the stored parking lots information. After the V2V matching process, for each matched EV as energy consumer, the achievable utility through the cooperative V2V charging will be checked whether to be larger than the utility when getting charged at a nearby charging station. If not, the corresponding matched EV pair will be marked as unmatched and put into the energy trading buffer again. Similarly, for each matched EV as energy provider, if the achievable utility through the cooperative V2V charging is not a positive value, the corresponding matched EV pair will also be marked as unmatched and put into the energy trading buffer again. If an EV as an energy consumer fails to be matched for more than m times, the data control center will feedback a failure-matched notice, which means it is a better option for the EV to get charged at the nearby charging stations. If a cooperative V2V charging deal is finally reached, the two involved EVs will perform the power transfer at a nearby parking lot selected by the data control center.

5.2.3.6 Performance Evaluation

Simulation results in [17] indicated that the cooperative V2V charging protocol with different V2V matching algorithms can effectively improve the utilities of the EVs. From Fig. 5.5, it can be clearly found that with the proposed cooperative V2V charging protocol, the utilities of EVs as energy consumers are improved significantly, leading to smarter and more effective charging behaviors. It can be also seen that EV_6^C has the same utility value with our proposed energy management protocol and the traditional EV charging protocol. This implies that for EV_6^C, cooperative V2V charging with other EVs as energy providers in the network cannot lead to a better utility than to get charged at the charging station. Therefore, it finally chooses to get charged at the nearest charging station based on the feedback decisions from the data control center.

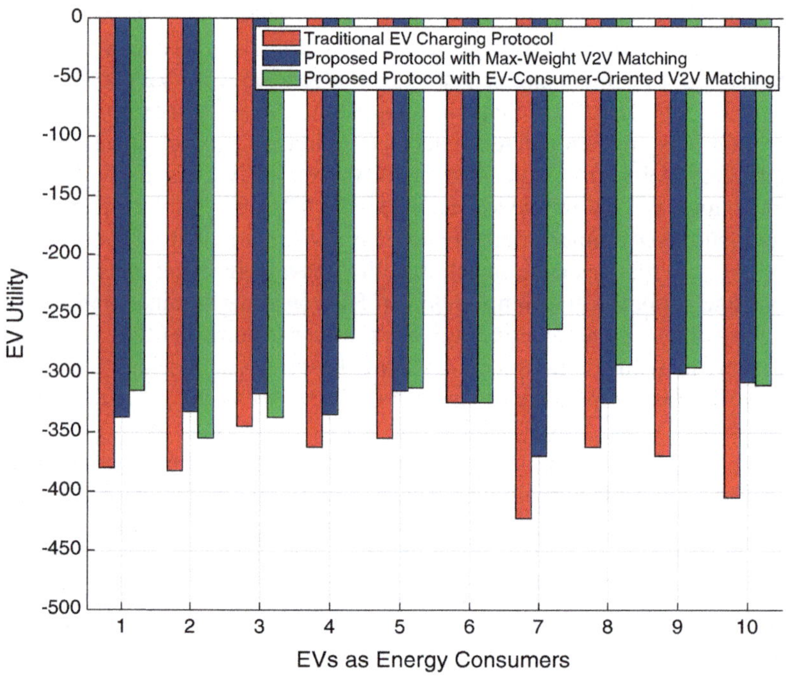

Fig. 5.5 Utility performance comparison for EVs as energy consumers ($N = K = 10$)

From Fig. 5.6, it can be seen that most EVs as energy providers can achieve a positive utility value, which makes the EVs that have extra power have an incentive to participate in the cooperative V2V charging process as energy providers. It can be also found that EV_7^P has zero utility value, which means EV_7^P does not find an effective partner for energy trading (i.e., unmatched) with the cooperative V2V charging protocol. There are two reasons resulting in this situation. First, EV_7^P cannot achieve a positive utility based on the V2V matching solutions and thus it prefers to be unmatched. Second, the matched partner of EV_7^P cannot achieve a better utility through cooperative V2V charging than to get charged at the charging station (e.g., EV_6^C in Fig. 5.5), and thus the matched partner leaves the matched relation, making EV_7^P also become unmatched. Also note that the utilities of EVs as energy providers with the EV-provider-oriented stable V2V matching algorithm are never smaller than those with the EV-consumer-oriented stable V2V matching algorithm. This is because the EV-provider-oriented stable V2V matching algorithm can always yield the EV-provider-optimal output.

In Fig. 5.7a, b, the energy consumption reduction performance of all the involved EVs through the cooperative V2V charging protocol with different V2V matching

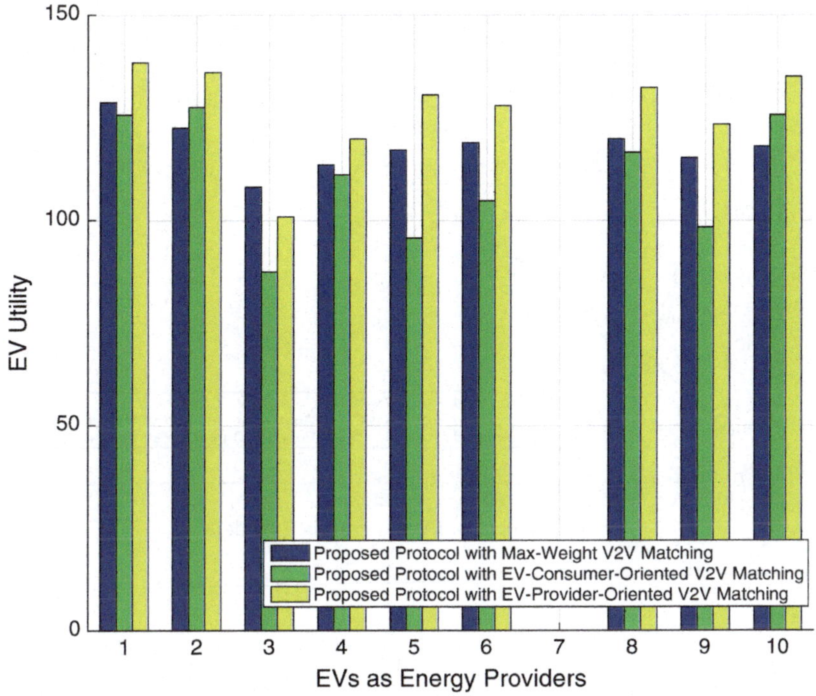

Fig. 5.6 Utility performance comparison for EVs as energy providers ($N = K = 10$)

algorithms is investigated compared with the traditional EV charging protocol where all the EVs as energy consumers choose to get charged at the nearest charging station. Actually, the energy consumption reduction is calculated as the network energy cost (i.e., the sum of energy cost of EVs as energy consumers and EVs as energy providers that finally participate in energy trading) difference between the cooperative V2V charging protocol and the traditional EV charging protocol. From Fig. 5.7a, b, it can be clearly found that the energy consumption of the involved EVs can be reduced effectively through the V2V charging protocol with all the three V2V matching algorithms. This leads to a more flexible and smarter energy management for the EV system. In addition, the max-weight V2V matching algorithm can always lead to best network performance but it does not take the individual rationality of the EVs into consideration, which means the obtained V2V matchings may be not stable if the EVs can make decisions based on their own utilities. Whereas the EV-consumer-oriented and EV-provider-oriented V2V matching algorithms can output EV-consumer-optimal and EV-provider-optimal stable V2V matchings with very low computational complexity, respectively.

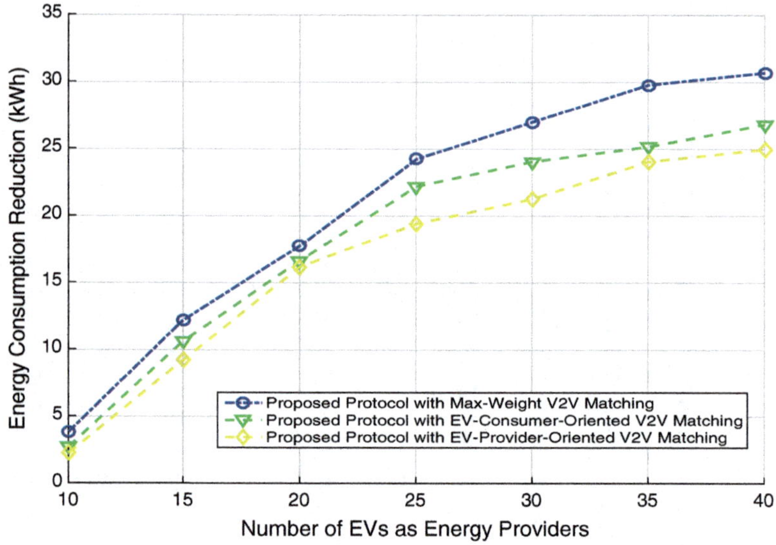

(a)

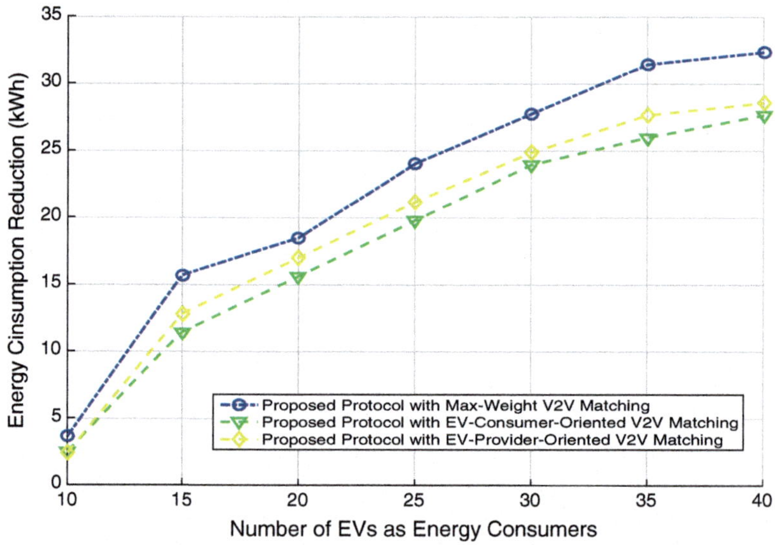

(b)

Fig. 5.7 Energy consumption reduction through the cooperative V2V charging protocol with different V2V matching algorithms. (**a**) $N = 10$, K changes. (**b**) $K = 10$, N changes

5.3 Distributed Data Storage

Recently emerging vehicular applications, including road safety, intelligent transportation, in-vehicle entertainment, self-driving, etc., lead to rapidly increasing vehicular communication requirements. This implies significant burden on the core networks as well as the road side units (RSUs), and consequently result in quality-of-service (QoS) degradation of vehicular users. In the literature, the concept of information-centric networking (ICN) has been proposed to shift the host-centric networking paradigm into a content-centric one [20]. In this context, in-network caching is a key technique to bring contents closer to users. By caching contents at the edge of networks, duplicated requests for cached contents could be largely reduced in core networks. In the wireless setting, caching in the device level has been demonstrated to improve the spatial spectrum efficiency by exploiting local wireless links in the literature [21]. This can also significantly alleviate the load of both core networks and the original content servers as well as improve user experiences. However, the burden on road side infrastructure in vehicular networks remains the same. Although diverse research has been carried out on this subject, the application of ICN in vehicular scenarios is quite limited. Most of such work focus on the RSU, and only a few utilize in-vehicle caching but with insufficient and impractical considerations on the vehicle mobility. Whereas in fact, mobility consists of one core obstacle against the realization of in-vehicle caching. In this section, we aim to explore and discuss the possibility and feasibility of in-vehicle caching, and provide an efficient in-vehicle caching framework based on dynamic distributed storage relay (D^2SR) mechanism.

5.3.1 *In-Vehicle Caching Framework*

To resolve the challenging high vehicle mobility issue, in [22], an in-vehicle caching (IV-Cache) framework was proposed based on an innovative integration of distributed storage with cached content relay facilitated by one-hop V2V links. At the foundation of this framework, a dynamic distributed in-vehicle storage system with satisfying data survival probability within a designated region of interest is provided, which is robust against vehicle mobility and volatile V2V communication links. Specifically, due to the high mobility, a vehicle with stored data could only physically reside in a region of interest for a short period of time, except for the situations of heavy traffic jams or congestions. To cope with the data loss caused by vehicles leaving the region of interest, data transfer regions are allocated at each entrance/exit of the region so that the stored data can be relayed from leaving vehicles to incoming vehicles via one-hop V2V links. As a result, the stored data can be retained within the region of interest.

However, such a data storage scheme is vulnerable to unreliable V2V links, which may lead to data loss during the data relay. This problem is particularly severe in high-mobility scenarios. Erasure coding is a commonly employed technique to accommodate unreliable storage nodes in distributed storage systems [23]. Inspired by this, structured redundancy by erasure codes, e.g., maximal distance separated (MDS) codes, is also introduced to deal with the vulnerability of the distributed storage scheme caused by unreliable V2V links to maintain the integrity of stored data even when portions of coded data blocks are lost. In [24], a vehicle-to-vehicle (V2V) distributed storage was proposed to resolve high mobility issue in vehicular networks. However, without considering the retrieval performance, such storage system cannot provide a robust storage service for the caching application. In order to facilitate the IV-cache framework, in [22], a dynamic distributed storage relay (D^2SR) mechanism was designed, which ensures the survival of cached contents with a high probability within the time duration of interest. From the perspective of content-centric networking, D^2SR ensures the survivability of a content in the designated region rather than its carrying vehicle. With the robust storage service provided by the D^2SR mechanism, an effective IV-caching placement scheme was further provided by accommodating the content popularity, where cached contents are first encoded by MDS codes and then allocated to vehicles in the designated region. Thus, the cached contents stored at vehicles can be maintained by the D^2SR mechanism. As a result, most requests for the cached contents can be served by the vehicles carrying the corresponding contents via V2V communications and data traffic on cached contents are offloaded from the RSUs to vehicles in the designated region.

5.3.2 *Dynamic Distributed Storage Relay (D^2SR) Mechanism*

Figure 5.8 illustrates a vehicular network in a designated bidirectional two-lane region with length $(L_c + 2L_t)$ and width D, as shown in Fig. 5.8. The designated region is covered by RSUs, via which vehicles in the region can reach the backend server by vehicle-to-infrastructure (V2I) links. The effective range of a V2V link between two vehicles is assumed to be R.

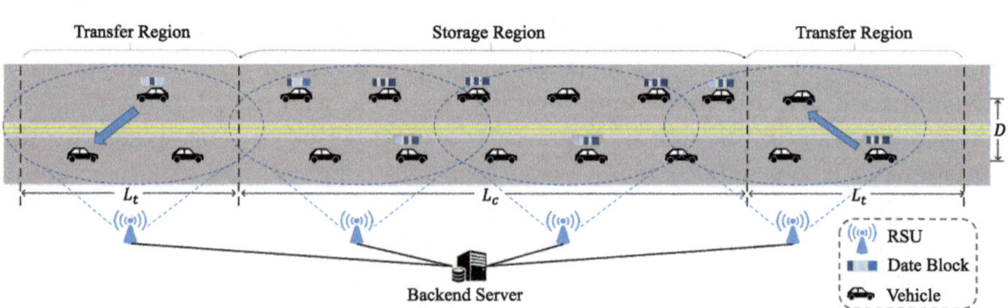

Fig. 5.8 Dynamic distributed storage relay (D^2SR) scheme overview

We aim to cache a popular content library at vehicles in this designated region. Hence, requests for the cached contents in the designated region could be served by vehicles associated with such contents via V2V links rather than by the RSUs so that the burden on RSUs could be alleviated. The cached content will be lost once its carrying vehicle moves out of the region, leaving the cached contents limited opportunity of being hit. This is a critical hurdle constraining the usefulness of in-vehicle caching. Hence, the IV-Caching scheme consists of two key components:

- *D^2SR:* A dynamic distributed storage relay system based on V2V communication is proposed to provide robust storage service in the designated region so that the contents cached at vehicles is no longer loss-prone.
- *IV-Caching Placement:* Atop the proposed vehicle-based distributed storage system, an IV-Caching placement scheme enhanced with content popularity is proposed to improve the network performance.

5.3.2.1 D²SR System Overview

The proposed D^2SR scheme is designed to ensure the survival of the stored data in the designated region with vehicles as storage nodes. We introduce redundancy to the stored data and employ storage relay via one-hop V2V links to prevent data loss due to the high mobility of vehicles. The proposed D^2SR system consists of three components as follows.

- *Storage Coding and Allocation:* Assume that the entire region is covered by RSUs and a backend server behind RSUs is employed to control and manage the D^2SR system. In the beginning, the backend server encodes a file with k data blocks into n ($n < N$) coded data blocks using an MDS code. Each coded data block is then allocated to one vehicle in the region through the V2I links. Hence, n coded data blocks of a file are stored at n vehicles initially, out of N denoting average total vehicle number in the storage region. The vehicle flow statistic model is based on the headway between two consecutive vehicles along the same direction [25].
- *Data Relay in Transfer Regions:* However, coded data blocks will be lost when their carrying vehicles leave the storage region. Data relay mechanism is proposed to ensure that the coded data blocks are retained within the storage region. The specific rules are as follows. Vehicles will start to attempt to relay their stored coded data blocks to vehicles they meet via one-hop V2V links, once they enter the transfer region from the storage region. A vehicle will terminate relaying attempts when it moves out of the transfer region or its transmission procedure is activated. The reason why a vehicle will not retry to relay its data block if the transmission fails is that excess transmission may bring interference to other relaying processes in the transfer region. As a result, each of the coded data blocks that is not successfully relayed will be considered as a loss in our proposed D^2SR scheme. Nevertheless, the MDS codes could tolerate up to $(n-k)$

coded data block losses and still guarantee the survival of the stored data in the storage region.

- *Data Retrieval in Storage Region:* When a vehicle enters the storage region, it first fetches the list of contents stored in the region from the backend server via RSUs. Based on MDS coding, k coded data blocks need to be collected by the vehicle for a successful file retrieval. Hence, when requesting the stored file in the region, one will first check its own storage and try to fetch coded data blocks from vehicles it meets through V2V links to collect sufficient coded data blocks. With the V2V data relay mechanism and the data redundancy employed, our proposed D^2SR system could provide a robust storage service for IV-Cache, which will be described next.

Due to unreliable V2V links, coded data blocks may not be fully transmitted within the short lifetime of V2V connections. The coded data block is considered as a loss in our storage system when data transfer fails in transfer regions. Therefore, the transfer failure probability of V2V links is first studied, based on which the performance of our proposed D^2SR system, including data survival time and storage capacity, will be investigated. In addition, a special setup of the proposed D^2SR system is discussed to further enhance the system robustness for the proposed IV-Caching system.

5.3.2.2 Coded Data Block Loss

The probability of one coded data block loss in transfer regions is first studied. It is obvious that if a vehicle with a coded data block does not meet any vehicles in the transfer region, such coded data block will be lost. Therefore, we first study the event that a vehicle does not meet an incoming vehicle in the opposite-direction lane in the transfer region (Not Meeting Probability).

Not Meeting Probability Suppose that the average speed of the vehicle flows in two opposed directions are v_0 and v_0', respectively. To study the not meeting probability, we first investigate two boundary cases about vehicle positions, which can ensure that a vehicle attempting to transfer data block could meet at least one incoming vehicle in the transfer region as follows:

1. When the vehicle with a coded data block is just leaving the storage region, an incoming vehicle is at position 1 as shown in Fig. 5.9;
2. when the vehicle with a coded data block is just leaving the transfer region, an incoming vehicle is at position $2'$. It is equivalent to the case that the incoming vehicle is at position 2 when the vehicle is about to leave the storage region.

Any cases between the above two cases will guarantee that the vehicle with coded data block can meet an incoming vehicle in transfer regions. Hence, the event of not meeting any incoming vehicle is equivalent to the event that when the vehicle with coded data block arrives at the entrance of the transfer region, no incoming vehicle exists in the yellow area shown in Fig. 5.9. Here, we assume the vehicle number in this area N_{vya} follows Poisson distribution with mean n_{vya}, i.e.,

$$P(N_{\text{vya}} = k) = \frac{n_{\text{vya}}^k}{k!} e^{-n_{\text{vya}}} (k = 0, 1, 2, \ldots), \tag{5.11}$$

based on the commonly used exponential distribution for headway modeling [25]. The average headway is denoted as μ, thus the average number of vehicles in this area is given by $n_{\text{vya}} = \left\lfloor \frac{L_t' + L_t}{\mu} \right\rfloor$. Therefore, not meeting probability is given by

$$P_{nm} = P(N_{\text{vya}} = 0) = e^{-\frac{\frac{v_0}{v_0'} L_t + L_t}{\mu}}. \tag{5.12}$$

Not Completely Transmitted Probability According to our data relay rules, only one transmission attempt is allowed in the transfer region. In fact, the key to determine coded data block transfer failure at one transmission attempt is to evaluate the distribution of transmission size E_t at a transmission attempt. Rician channel is adopted to model the light-of-sight (LOS) small-scale fading of V2V channel [26]. Therefore, the channel capacity between two vehicles is given by

$$c = W \log_2(1 + \eta |h|^2 d^{-4}), \tag{5.13}$$

where d is the distance between the two vehicles; η is the signal-to-noise ratio (SNR) at transmitters; and W denotes the bandwidth. The Rician channel gain h is given as

$$h = \sqrt{\frac{\kappa}{\kappa + 1}} e^{j\theta} + \sqrt{\frac{1}{\kappa + 1}} \omega, \tag{5.14}$$

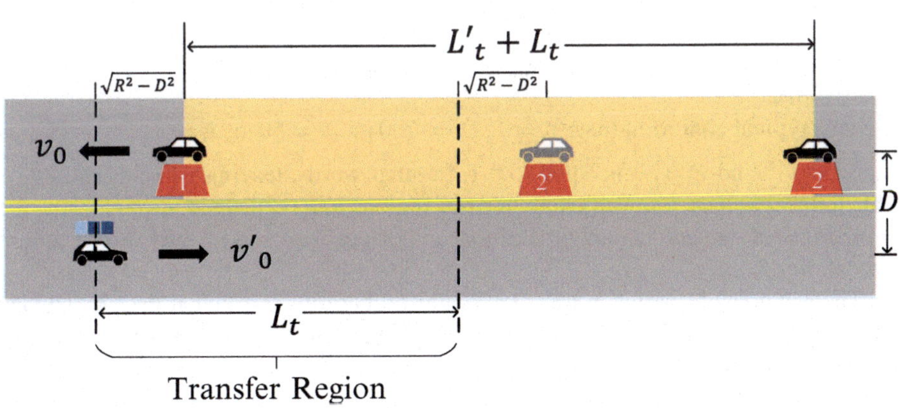

Fig. 5.9 Vehicle meeting model

where θ is a random variable uniformly distributed on $[0, 2\pi]$; ω denotes a complex Gaussian random variable with unit variance and zero mean and κ is the ratio of the energy in the LOS path to the energy in scattered paths. Hence, the transmission size of one V2V connection is given in the following theorem.

Theorem 5.1 *The transmission size E_t during a V2V connection is a non-linear function of the non-central chi-square distributed variable $X = 2(\kappa + 1)|h|^2$, i.e.,*

$$E_t = q\left[\ln\left(\frac{X}{2(\kappa + 1)}\right) + f\right]. \tag{5.15}$$

where q and f are dependent on (1) *V2V link bandwidth,* (2) *vehicle speeds (v and v'), and* (3) *V2V communication range R.*

Based on results in [27], the approximated cumulative distribution function of X can be obtained as

$$F_X(x) \approx \Phi\left\{\frac{(\frac{x}{2+2\kappa})^r - 1 - r\alpha[r-1-0.5(2-r)m\alpha]}{r\sqrt{2\alpha}(1 + 0.5m\alpha)}\right\}, \tag{5.16}$$

where $r = 1 - \frac{2}{3}\frac{(1+\kappa)(1+3\kappa)}{(1+2\kappa)^2}$, $\alpha = \frac{2+4\kappa}{(2+2\kappa)^2}$, $m = (r - 1)(1 - 3r)$, and $\Phi(\cdot)$ denotes the cumulative distribution function of standard normal distribution. Given the size of coded data blocks E, the transfer failure occurs when the transmission size via the V2V link cannot fulfill the demand of the coded data block transmission, i.e., $E_t < E$. By (5.15), the transfer failure is equivalent to $X < 2(\kappa + 1)e^{(\frac{E}{q} - f)}$. Therefore, the transfer failure probability $P_{tf}(E)$ can be obtained as

$$P_{tf}(E) = F_X\left(2(\kappa + 1)e^{(\frac{E}{q} - f)}\right). \tag{5.17}$$

Hence, the probability of coded data block loss in the transfer region is

$$P_{l1} = P_{nm} + (1 - P_{nm}) \times P_{tf}.$$

Given a typical condition that $\frac{v_0}{v_0'} = 1$, $L_t = 200\,\text{m}$, $\mu = 50\,\text{m}$, $E = 300\,\text{Mb}$, P_{nm} is equal to e^{-8} and $P_{tf}(E)$ is equal to 0.1. In other words, leaving vehicles should be able to meet at least one vehicle in transfer regions in a typical situation. Hence, the probability of one coded data block loss can be approximated by the transfer failure probability, i.e., $P_{l1} \approx P_{tf}$.

The relationship between transfer failure probability and coded data block size given different vehicle speeds is shown in Fig. 5.10. It can be observed that the transfer failure probability with the size of coded data blocks $E = 250\,\text{Mb}$ approaches zero when the speed of vehicles is less than 15 m/s. As the speed of vehicles increases, the connection lifetime of V2V links will be curtailed, which in turn increases the transfer failure probability. Also, the transfer failure probability

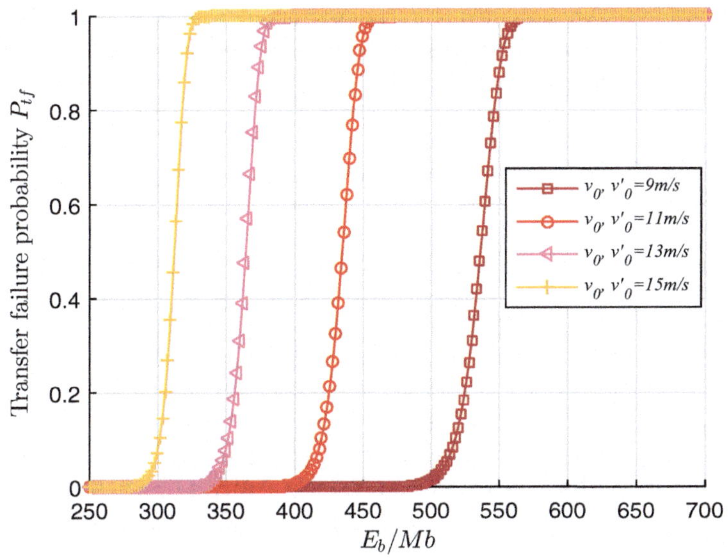

Fig. 5.10 The transfer failure probability and the size of coded data blocks with different vehicles speed

rises as expected when the size of coded data blocks increases. Hence, the transfer failure probability could be extremely low when the data block size is less than a threshold, e.g., $E_b < 300$ Mb when $v_0 = v_0' = 11$ m/s. This observation inspires the storage design for IV-Caching.

5.3.2.3 File Survival Time and Storage Capacity

To successfully reconstruct the original file, at least k coded data blocks are required in the storage region. For each coded data block, the probability of coded data block loss after Y transfers in the transfer regions can be obtained as

$$P(Y = y) = (1 - P_{tf}(E))^{(y-1)} P_{tf}(E). \tag{5.18}$$

For each coded data block, the time interval between two consecutive relays could be approximated as $t_0 = \frac{L_c}{2v_0} + \frac{L_c}{2v_0'}$. Thus, we approximate the survival time of a coded data block as the number of time intervals between two consecutive relays $t = Yt_0$. As a result, the probability of coded data block loss in terms of its size and survival time is

$$P_l(E, t) = P(Y \le y) = \sum_{n=1}^{y} (1 - P_{tf}(E))^{(n-1)} P_{tf}(E). \tag{5.19}$$

Coded data blocks are assumed to be homogeneous in our proposed data storage scheme. Hence, the survival of a file is equivalent to any k coded data blocks still existing in the storage region. Hence, the file survival probability can be obtained as

$$P_{su}(t, E, n, k) = \sum_{j=0}^{n-k} C_n^j P_l(E, t)^j [1 - P_l(E, t)]^{n-j}. \tag{5.20}$$

As a result, the survival time of a file T_{su} in the storage region is defined based on the survival probability as

$$T_{su} = \{t | P_{su}(t, E, n, k) \geq P_0, \ 0 < t \leq T_{su}\}, \tag{5.21}$$

where P_0 denotes the threshold of file survival probability.

If all vehicles in the storage region are used, with the same coding ratio, the storage capacity of this system is $C(N, E, k) = \frac{k}{n} N E$, $(1 \leq k \leq n \leq N)$, where $N = \left\lfloor \frac{L_c}{\mu_0} \right\rfloor + \left\lfloor \frac{L_c}{\mu_0'} \right\rfloor$ is the vehicle number in the storage region. It is worth noting that the storage capacity in this paper refers to the maximum amount of the original file before MDS coding instead of the sum of all available storage spaces. With the desired file survival time T_{su} under the file survival probability threshold P_0, an optimization problem to maximize the storage capacity can be formulated as

$$\begin{aligned} \underset{E,k}{\text{maximize}} \quad & C(N, E, k) \\ \text{subject to} \quad & P_{su}(T_{su}, E, N, k) \geq P_0 \\ & E > 0 \\ & 1 \leq k \leq N, k \in \mathrm{Z} \end{aligned} \tag{5.22}$$

Noted that at most N coded data blocks can be stored in the proposed D^{2}SR. Increasing either k or E will enlarge the overall storage capacity, at the cost of degraded file survival probability for a given T_{su}. Due to the monotonicity of P_{su} with respect to E and the bounded integer k (i.e., $k \leq N$), the optimal storage capacity can be achieved by solving (5.22) via exhaustive searching.

With the speed of vehicles $v_0 = v_0' = 15$ m/s and the size of coded data blocks $E = 280$ Mb, it is observed that more redundancy could help files survive longer as shown in Fig. 5.11, where The file survival time is defined as the length of the time interval in which the file survival probability remains at least P_0. However, the capacity of our storage system reduces when more redundancy is introduced in the proposed D^{2}SR scheme. Hence, another tradeoff can be clearly observed that the storage capacity increases along with the enlarging of coded data block size while the file survival time degrades.

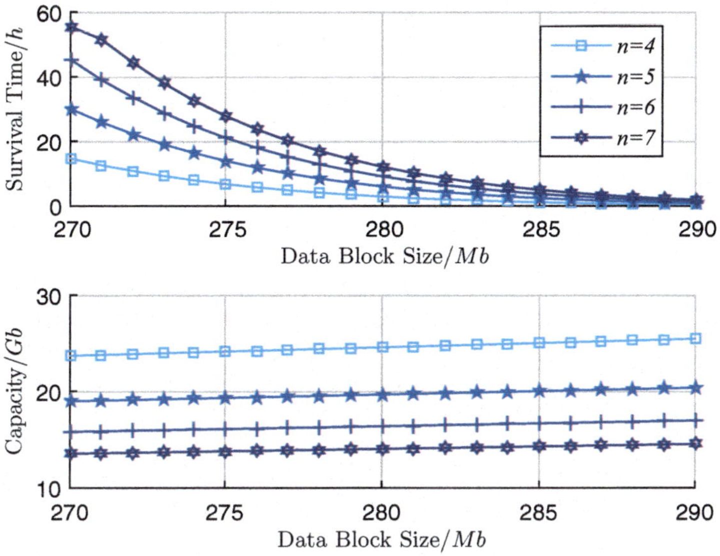

Fig. 5.11 File survival time, storage capacity and the size of coded data blocks with different coded data block number and $L_c = 3000$ m

5.3.2.4 D^{2}SR Setup for IV-Caching

The proposed general D^{2}SR is set up in order to provide a robust storage service for the IV-Cache, through which the data block loss in the storage region during the time of interest is significantly reduced. As illustrated in (5.17) and (5.20), the size of data blocks stored in vehicles is the key to the survival of a file in the storage region. Hence, the size of data blocks is subject to a limit, i.e., $E_b \leq E_{b,max}$, in order to ensure negligible data loss during the time of interest. According to (5.20), the no-loss probability can be obtained as

$$P_{nl}(t, E, N) = P_{su}(t, E, N, N) = [1 - P_l(E, t)]^N. \tag{5.23}$$

Similar to (5.21), the no-loss time T_{nl} is defined as follows,

$$T_{nl} = \{t | P_{nl} \geq P_0', \ 0 < t \leq T_{nl}\}, \tag{5.24}$$

where P_0' denotes the threshold probability. In other words, we aim to obtain the maximum per-vehicle storage size by a threshold P_0' to ensure the survival of cached files within the time span T_{nl} with high probability. According to (5.17), the threshold $E_{b,max}$ can be obtained as

$$E_{b,max} = qf + \frac{q}{r}\ln\left(1 + r\alpha\left[r - 1 - \left(1 - \frac{r}{2}\right)\alpha\right] - r\sqrt{2\alpha}\,(4 + 2m\alpha)\right), \quad (5.25)$$

where r and α represent the same parameters as in (5.16).

5.3.3 *IV-Caching via D²SR*

In-network caching is the key technology to reduce the duplicated requests for the same contents in the network. In a wireless setup, in-network or in-device caching could not only curtail the burden on the wireless infrastructure, but also boost the frequency spatial reuse, as multiple copies of requested files might be simultaneously available at different nodes in the network.

5.3.3.1 IV-Caching Placement

With the robust storage service provided by D^2SR, the IV-Caching placement scheme is designed to accommodate the content library into the proposed D^2SR system with the content popularity considered. Specifically, an N_f-file content library has the popularity information, $\mathcal{F}=\{F_1, \cdots, F_{N_f}\}$, where F_i denotes the i-th most popular file. The request probability distribution over the library is modeled as a Zipf distribution [28] as follows,

$$f_i(\gamma_r) = \frac{1/i^{\gamma_r}}{\sum_{j=1}^{N_f} 1/j^{\gamma_r}}, \; i \in \{1, 2, \cdots, N_f\}, \quad (5.26)$$

where the exponent γ_r is the parameter characterizing users' requests for the library. In addition, each file in the library is assumed to have the same size E_f [26, 29, 30]. Each vehicle in the designated region will be utilized as a storage node with the same storage capacity $E_{b,max}$. With the storage demands by the library $\mathcal{F}$, the key problem is how to place the library into the D^2SR system with the content popularity considered. Intuitively, the file with higher popularity should have higher availability in the network. However, due to the limited storage capacity of the storage system, increasing the availability of popular files means reducing the availability of relatively unpopular files, which may lower the overall average retrieval probability.

In order to achieve the maximum overall average retrieval probability, a cache placement scheme is proposed by tuning the data redundancy based on the Zipf distribution, inspired by the one in [28]. Two phases in the cache or storage placement process, namely the coding and allocation phases, are summarized as follows.

1. *Coding*: In the beginning, each file in the library is divided into k data chunks[1] and the size of each data chunk is $E_c = E_f/k$. Specifically, the case $k = 1$ means that files in the library are not divided at all. Thus, the k data chunks of file F_i are individually encoded into n_i ($k \leq n_i$) coded data chunks by an MDS code, where the number of coded data chunks is dependent on the rank i of file F_i. In other words, a file with different popularity will have different redundancy level adjusted by a controllable parameter in the system. Hence, the number of coded data chunks n_i of file F_i is

$$n_i = \max\left\{k, \left\lfloor N \times n_c \times f_i(\gamma_c) + \frac{1}{2}\right\rfloor\right\}, \tag{5.27}$$

   where n_c is the number of data chunks a vehicle could hold with no data loss almost ensured, i.e., $n_c = \lfloor E_{b,max}/E_c \rfloor$, $N \times n_c$ denotes the total number of data chunks the D^{2}SR system can hold and $f_i(\gamma_c)$ is the probability mass function of a Zipf distribution with a controllable parameter γ_c. The Zipf distribution $f_i(\gamma_c)$ controls the data redundancy of files in the library based on their popularity rank with a tunable Zipf distribution parameter γ_c.
2. *Allocation*: The total $\sum_i^{N_f} n_i$ coded data chunks in the library is first concatenated and then randomly shuffled at the backend server. Every n_c randomly coded data chunks are combined as a coded data block, which will be allocated to a vehicle in the region via RSU-to-vehicle (R2V) links. Hence, the coded data chunks of N_f files randomly exist in N vehicles, each of which holds one data block with n_c data chunks.

The proposed cache placement scheme is also illustrated in Fig. 5.12.

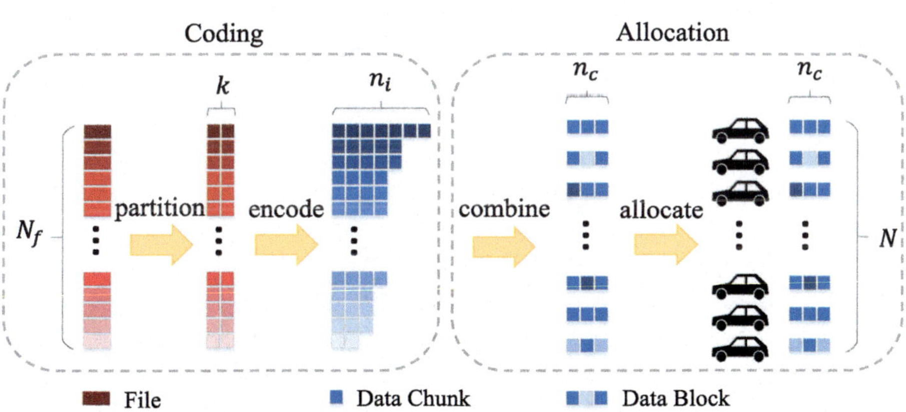

Fig. 5.12 Cache placement scheme

[1]The data chunk is defined as a finer data unit in this paper, compared with the data block. In other words, a data block could comprise many data chunks.

5.3.3.2 V2V Link Establishment

The cached data retrieval depends on the number of coded data chunks of the desired file that can be collected in the storage region. As the maximum data block size that a vehicle can carry is restricted to $E_{b,max}$, the probability of successful data transmission of one data block for cached data retrieval is very high, not to mention that the size of demanded coded data chunks might be much smaller than that of a data block. Therefore, it is assumed that every data transmission could be completed almost surely under such a setting once the V2V link is established.

However, a V2V link establishment is not only constrained by the spatial distance between two vehicles, but also by the wireless medium competition among multiple V2V links. The communication range R is employed to define the maximum distance of two vehicles, within which a V2V link can be established. In addition, we design an extra protective distance δR in order to ensure that each established V2V link is interference free. In other words, two vehicles could establish a V2V link only if no V2V link exists within their $(1+\delta)R$ neighbors. Hence, a general case is considered that a V2V link of two vehicles could be established, as shown in Fig. 5.13, where these two vehicles are surrounded by four neighbors. The following

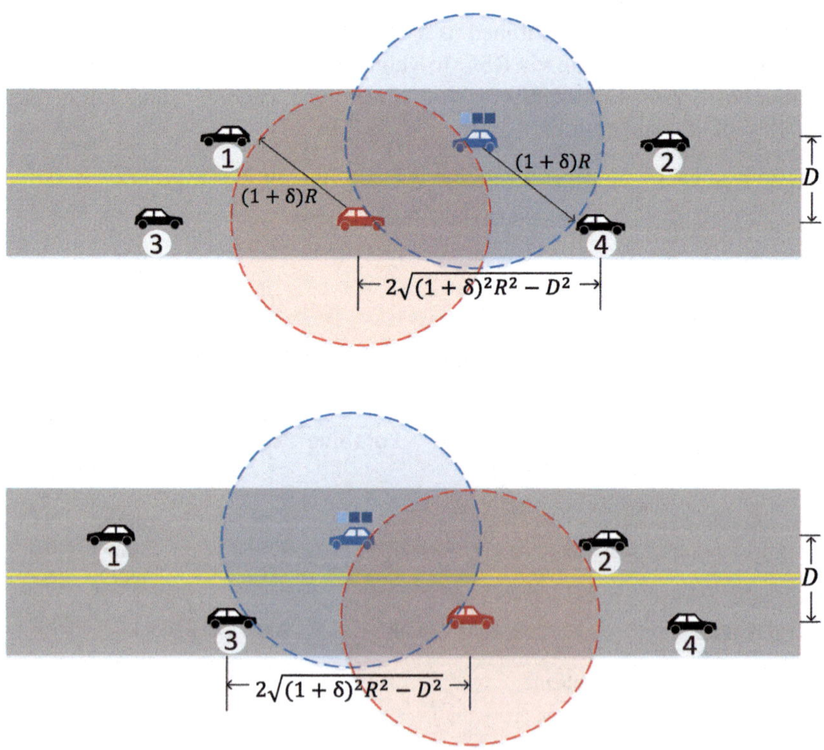

Fig. 5.13 V2V interference model

two conditions are critical for a successful V2V link establishment. First, the neighbors within $(1+\delta)R$ distance are not occupying the medium. Here, a parameter $p_c \in [0, p_{c,max}]$ is employed to characterize the medium occupying probability. Second, even when neighbors are accessing the medium, the V2V link could be established as long as the distance to their neighbors is beyond $(1+\delta)R$. Therefore, the headway between two consecutive vehicles d_n should satisfy

$$d_n > \max\left\{2\sqrt{(1+\delta)^2R^2 - D^2}, (1+\delta)R\right\}.$$

Based on the two-lane Gaussian distributed headways with expectation μ, μ' and standard deviation σ, σ', as a result, the probability of a successful V2V link establishment is obtained as

$$p_{link}(d_n) = \left(1-p_c+p_c Q\left(\frac{d_n-\mu}{\sigma}\right)\right)^2 \left(1-p_c+p_c Q\left(\frac{d_n-\mu'}{\sigma'}\right)\right)^2, \tag{5.28}$$

where $Q(\cdot)$ is the tail distribution function of the standard normal distribution.

5.3.3.3 Cached File Retrieval

Besides the establishment of V2V links, the number of desired coded data chunks a vehicle can collect is another critical factor. The probability that a coded data chunk belongs to the desired file F_i is modeled as $p_{cf,i} = \frac{n_i}{Nn_c}$. Furthermore, the number of the vehicles that a file-requesting vehicle can meet in the storage region is also important for a successful cached data retrieval. Here, the average number of vehicles in the opposite lane that one vehicle could meet in the storage region is denoted as

$$N_{\text{meet}} = \left\lfloor \frac{\frac{v_0}{v_0'}L_c + L_c}{\mu^{(\prime)}} \right\rfloor. \tag{5.29}$$

Again, $\mu^{(\prime)}$ is the mean of Gaussian distributed headways.

$$p_{b,i}(n_b) = \begin{cases} \mathrm{P}(B(p_{cf,i}, n_c) \geq k), & n_b = 0 \\ \sum_{j=0}^{n_b-1}\Big[\mathrm{P}\left(B\left(p_{link}, n_b-1\right) = j\right)\sum_{z=0}^{k-1}\mathrm{P}\left(B\left(p_{cf,i}, (j+1)\, n_c\right) = z\right) \\ \mathrm{P}\left(B\left(p_{cf,i}, n_c\right) \geq k-z\right) p_{link}\Big], & n_b \in \{1, \cdots, N_{\text{meet}}\} \end{cases} \tag{5.30}$$

As a result, a $(N_{\text{meet}} + 1) \times 1$ vector Ψ is employed to summarize the results of V2V link establishments of a file-request vehicle in the storage region, where its j-th element denotes the link establishment status when it meets the j-th vehicle,

$$\psi_j = \begin{cases} 1 & \text{link established,} \\ 0 & \text{otherwise.} \end{cases}$$

Also, another $(N_{\text{meet}}+1) \times 1$ vector Λ^i is utilized to record the number of the coded data chunks of the desired file F_i in each vehicle it meets, i.e.,

$$\lambda^i_j = z, z \in [0, 1, \cdots, n_c].$$

Hence, the total number of the coded data chunks of file F_i that a vehicle can collect in the storage region is

$$N_{\text{collect},i} = \Psi^T \Lambda^i, \tag{5.31}$$

where $\{\cdot\}^T$ denotes the vector transpose.

It is worth noting that ψ_j and λ^i_j are random variables employed for cached file retrieval analysis. First, the probability that a vehicle with the request for file F_i can successfully retrieve the file after meeting $n_b \in \{0, 1, \cdots, N_{\text{meet}}\}$ vehicles is calculated as shown in (5.30). Specifically, $n_b = 0$ means that the vehicle requesting F_i can successfully retrieve F_i from its own cache. Since a successful cached file retrieval requires at least k coded data chunks, the successful probability when $n_b = 0$ can be calculated as $p_{b,i}(n_b = 0) = \mathrm{P}\left(B(p_{cf,i}, n_c) \geq k\right)$, where $B(p, n)$ denotes the binomial random variable with parameters p and n.

When $1 \leq n_b \leq N_{\text{meet}}$, its underlying assumption is that the coded data chunks collected from $(n_b - 1)$ vehicles are insufficient for a successful cached data reconstruction. Hence, we study the case of j total successful link establishments out of $(n_b - 1)$ vehicles, whose probability can be obtained as $\mathrm{P}\left(B\left(p_{link}, n_b - 1\right) = j\right)$. Thus, the total number of coded data chunks that can be collected is $(j + 1) \times n_c$, where n_c is the number of the chunks stored in one vehicle. Under the assumption that j links established after meeting $(n_b - 1)$ vehicles, the probability of the successful retrieval of cached file F_i when meeting the n_b-th vehicle is

$$\sum_{z=0}^{k-1} \mathrm{P}\left(B\left(p_{cf,i}, (j+1)\, n_c\right) = z\right) \mathrm{P}\left(B\left(p_{cf,i}, n_c\right) \geq k - z\right) p_{link}.$$

Hence, the retrieval probability of file F_i after meeting n_b vehicles in the storage region can be obtained by (5.30).

5.3.3.4 Retrieval Performance Analysis

With the analysis of cached file retrieval, key network performance metrics are investigated as follows.

- *Cached File Retrieval Outage* Cached file retrieval outage is defined as cached file retrieval failure in the storage region. In other words, the vehicle cannot collect k desired coded data chunks, i.e, $\Psi^T \Lambda^i < k$. Hence, the cached file F_i retrieval outage can be calculated as

$$P_{out,i} = 1 - \sum_{n_b=0}^{N_{\text{meet}}} p_{b,i}(n_b). \tag{5.32}$$

  Hence, the average outage probability based on the Zipf distribution is obtained as

$$P_{out} = \sum_{i=1}^{N_f} f_i(\gamma_r) P_{out,i}. \tag{5.33}$$

- *Average Retrieval Delay* The average retrieval delay $\tau_{avg,i}$ is defined as the expected time span that a vehicle starts collecting the coded data chunks of file F_i until it can successfully reconstruct the file. When a retrieval outage occurs, the entire file will be transmitted from the RSU with the time cost τ_{inf}. Hence, the average retrieval delay of file F_i based on cached file retrieval analysis (5.30) can be obtained as

$$\tau_{avg,i} = \left(\tau_{inf} + \frac{\mu N_{\text{meet}}}{v_0 + v_0'} \right) P_{out,i} + \sum_{n_b=0}^{N_{\text{meet}}} \frac{\mu n_b}{v_0 + v_0'} p_{b,i}(n_b). \tag{5.34}$$

  As a result, the average retrieval delay of the library is

$$\tau_{avg} = \sum_{i=1}^{N_f} f_i(\gamma_r) \tau_{avg,i}. \tag{5.35}$$

Figure 5.14a shows the average delay and the file caching coefficient γ_c in (5.27) with different file request coefficient γ_r. It can be observed that when the file caching coefficient γ_c increases, the average delay first decreases and then increases. The optimal γ_c in terms of average retrieval delay can be obtained by the line search on (5.35) with respect to γ_c. For instance, when $\gamma_r = 1$, the average delay reaches minimum when $\gamma_c = 0.48$. Besides, it can be observed that the optimal γ_c increases when γ_r grows. The γ_r characterizes the level of unbalance of file requests over the library. Hence, to accommodate severer request unbalance, more storage space should be allocated to files with higher popularity. Consequently, the optimal γ_c in (5.27) increases. In addition, the average outage probability and the file caching coefficient γ_c with different file request coefficient γ_r are also compared as shown in Fig. 5.14b. The behavior of γ_c with respect to the retrieval outage is similar to that with respect to the retrieval delay. However, it can be clearly noticed that

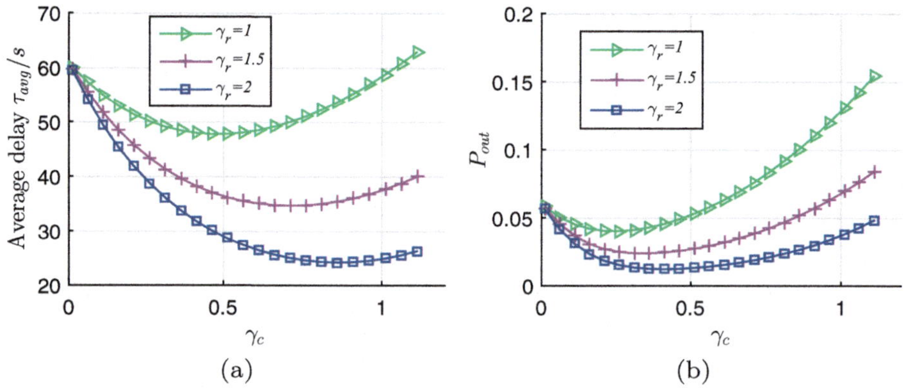

Fig. 5.14 Average delay and average outage probability versus file caching coefficient with different file request coefficient. (**a**) Average delay. (**b**) Average outage probability

the γ_c that optimizes the retrieval outage differs from the one that optimizes the retrieval delay, e.g., the γ_c that optimizes the retrieval outage with $\gamma_r = 1$ is 0.24, while the one that yields the lowest retrieval delay is 0.48. Furthermore, the retrieval delay and retrieval outage with respect to the file popularity ranks are illustrated in Fig. 5.15a, b, respectively. The retrieval delay of files with a higher popularity rank is significantly less than the one with a lower popularity rank. Similar behavior can also be observed in terms of the retrieval outage. However, the performance gain by the optimal retrieval delay is not as large as the one by the optimal retrieval outage. This might be caused by the delay penalty of fetching the desired file from RSUs when outages occur. Overall, the significantly lower retrieval outage probability shown in Figs. 5.14b and 5.15b suggests the successful data traffic offloading effect of the proposed IV-Caching scheme based on the D^2SR system. Figure 5.16a shows the average retrieval delay behaviors with respect to the file library size with different file sizes. It can be observed that the average delay increases with the file library size and the file size. Specifically, a vehicle in the proposed IV-Caching system can fetch a 160 Mb file from a 20-file library within 50 s. Similarly, the retrieval outage behaviors are illustrated in Fig. 5.16b, which suggests that 95% of requests for cached contents can be served by the IV-Caching system on average.

5.4 Physical Layer Security for VCN

Traditionally, the privacy of the confidential and private information is guaranteed by high layer cryptographic methods, which will bring significant communication overhead for secret keys generation and distribution due to the open and distributed accessing feature and the ever-increasing number of vehicles. Physical layer security

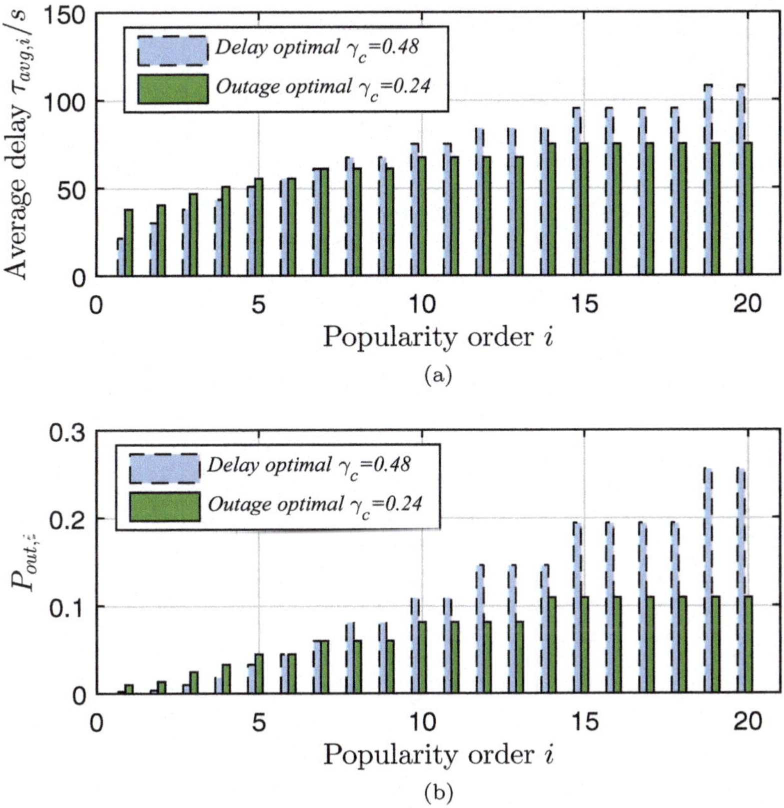

Fig. 5.15 Average delay of F_i and average outage probability of F_i versus popularity order with different file caching coefficient. (**a**) Average delay of F_i. (**b**) Average outage probability of F_i

[31–33] has been regarded as a promising technique to combat eavesdropping by exploiting the characteristics of wireless channels and signal processing techniques, and is able to provide perfect information-theoretic secrecy for wireless communications without relying on a secret key. The introduction of physical layer security techniques in V2X networks can effectively and efficiently secure V2X communications, and is compatible with most existing high-layer security protocols, which can further protect the confidential messages and individual information away from eavesdropping. Currently, many interesting physical layer security problems in vehicular networks are still open challenges. In this section, we first introduce some secrecy performance metrics as well as the challenges for physical layer security design in VCN, and then discuss possible applications of physical layer security to enhance the secrecy performance of VCN.

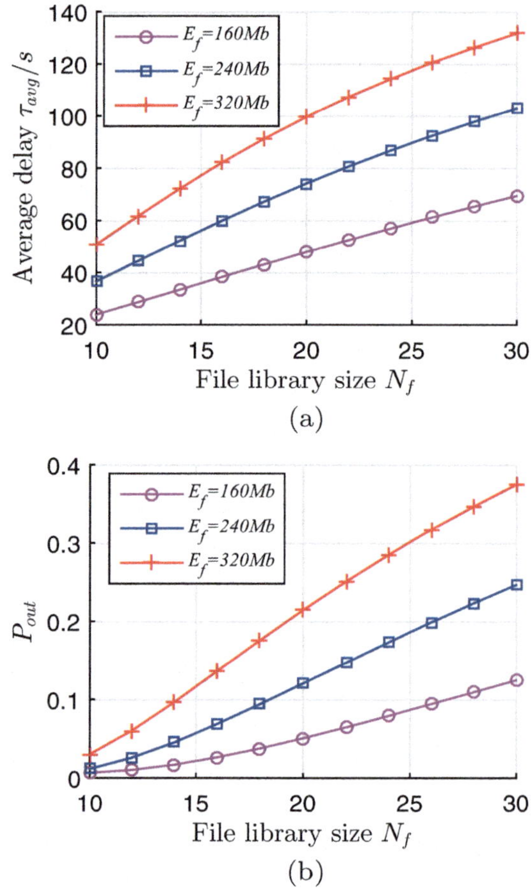

Fig. 5.16 Average delay and average outage probability versus file library size with different file sizes: (**a**) Average delay. (**b**) Average outage probability

5.4.1 Secrecy Performance Metrics

In order to effectively exploit physical layer security techniques to secure V2X communications, we introduce three information-theoretic secrecy performance metrics [34], including secrecy capacity, secrecy outage probability, and secrecy throughput, which can be used to evaluate physical layer security design and optimization in vehicular networks.

5.4.1.1 Secrecy Capacity

Secrecy capacity is the most fundamental evaluation metric of physical layer security. Introduced by Wyner in his pioneering work, when the wiretap channel is a degraded version of the main channel, the legitimate source and the intended destination can exchange perfectly secure messages at a positive rate, while the eavesdropper learns almost nothing about the confidential messages from its observations. This positive achievable information rate is termed *secrecy rate*, and the maximum secrecy rate from the source to the destination is defined as *secrecy capacity*. For the basic wiretap channel, suppose that the source input is X and the channel outputs at the destination and the eavesdropper are Y and Z, respectively. Then, the secrecy capacity C^s can be expressed by $C^s = \max\,[I(X;Y) - I(X;Z)]$, where the maximum is achieved through all possible input distributions (leading to a non-convex optimization problem in most cases), and $I(X;Y)$ and $I(X;Z)$ denote the mutual information between X and Y, and X and Z, respectively. To evaluate the secrecy performance more conveniently and computation affordably, in some cases we can simply use the Gaussian signal for X and regard the instantaneous wireless channel as a fixed one based on the obtained channel state information (CSI). Therefore, the achievable secrecy rate, which is a lower bound of the secrecy capacity, can be given as the difference between the achievable rates of the main channel and the wiretap channel with Gaussian codebook, that is, $C^s = \left(C_{\text{main}} - C_{\text{wiretap}}\right)^+$, where $(x)^+$ means $\max(x, 0)$. This simplified secrecy performance metric is widely used in many physical layer security based signal design and optimization problems, such as the precoding design for cooperative relaying, the power control for cooperative jamming, and the resource allocation optimization.

5.4.1.2 Secrecy Outage Probability

Secrecy outage probability is mostly used to evaluate the secrecy performance over wireless fading channels, which is defined as the likelihood that the instantaneous secrecy rate R^s is lower than a pre-defined threshold ϵ for a particular fading distribution, that is, $\mathcal{P}_{\text{out}} = \Pr(R^s < \epsilon),\ \epsilon > 0$. This indicates that perfect secrecy at a transmission rate R^s can be guaranteed by a probability $1 - \mathcal{P}_{\text{out}}$ under fading channels.

5.4.1.3 Secrecy Throughput

In practical networks, the achievable secrecy rate R^s can be adjusted adaptively according to the instantaneous CSI of the main channel to maintain a required outage probability. Hence, *secrecy throughput* can also be employed to evaluate the secrecy performance, which is defined as the average achievable secrecy rate over all channel realizations subject to a pre-determined secrecy outage probability. In network secrecy performance optimization, secrecy throughput has become an essential optimization objective for physical layer security design.

5.4.2 Challenges for Physical Layer Security Design in Vehicular Networks

Physical layer security has been regarded as a promising technique to combat eavesdropping and provide perfect information-theoretic secrecy for wireless communications. The introduction of physical layer security techniques in vehicular networks can effectively and efficiently secure V2X communications, and is compatible with most current high-layer security protocols, which is able to further enhance the network security. However, due to the characteristics of vehicular networks, it will be more challenging for feasible and effective physical layer security design therein. How to address the challenges significantly affects the efficiency in improving the network secrecy performance by exploiting physical layer security techniques.

5.4.2.1 Complex Heterogeneous Network

In vehicular networks, there are various types of V2X communications co-existing, including vehicle-to-infrastructure (V2I) communications, vehicle-to-vehicle (V2V) communications, vehicle-to-pedestrian (V2P) communications, and vehicle-to-cloud (V2C) communications, which results in a more complex network topology. Moreover, in such a heterogeneous network, the interference scenario becomes more complicated. This increases the difficulty and complexity in effective physical layer security design and optimization. How to guarantee the feasibility and efficiency of physical layer security techniques in such a complex heterogeneous network is a big challenge.

5.4.2.2 Multi-hop Data Transmission

Different from many other centralized networks, in vehicular networks, data transmissions among vehicles are usually performed in a multi-hop manner. This increases the risk of information leakage since the confidential data have to be transmitted over multiple relays (may be untrusted ones) and multi-hop wireless links. During the information routing processing, it is difficult to guarantee the perfect secrecy of the confidential data.

5.4.2.3 High Mobility and Imperfect CSI

Prior knowledge of the CSI of the main and wiretap channels is critical for the selection of secrecy performance metrics as well as the optimization of physical layer security design. For example, to maximize the achievable secrecy rate requires both instantaneous CSI of the main channel and the wiretap channel. However, due to the high mobility of vehicles, V2X communications experience fast time-variant channels, leading to severe imperfect CSI acquisition at the sources or the centralized coordinators and thus degraded achievable secrecy performance. Moreover, the network topology is also changing fast due to the high mobility of vehicles therein, which increases the difficulty of cooperative secrecy design (e.g., cooperative relaying and cooperative jamming) among vehicles.

5.4.3 *Enhance Secrecy via Cooperation*

According to the secrecy capacity defined above, the secrecy performance is actually determined by the superiority of the main channel to the wiretap channel. User cooperation have been regarded as an effective means to guarantee or improve this superiority by increasing the quality of the main channel and/or degrading the quality of the wiretap channel. Hence, in the literature, many cooperative signal designs and multiple antennas schemes have been employed in different wireless networks to improve the physical layer security performance [32–36]. Regarding the vehicular network, which is a complex heterogeneous network and includes various communication types, it will be more challenging for network optimization and resource scheduling for V2X communications. But at the meantime, this also provides a potential opportunity for different V2X communication links to enhance their secrecy performance via effective and efficient cooperation.

Referring to the literature, many cooperative secrecy schemes can be employed in vehicular networks to enhance the secrecy performance of V2X communications, including cooperative MIMO, cooperative relaying, cooperative jamming, and hybrid cooperative secrecy scheme, as illustrated in Fig. 5.17.

5.4.3.1 Cooperative MIMO

In vehicular networks, vehicle platooning has become a trend where a group of vehicles move in a closely linked manner to form a platoon for intelligent cooperative driving and information sharing. This actually provides a ready vehicle formation to perform distributed MIMO among vehicles within the same platoon. Distributed MIMO through vehicle platooning can effectively enhance the robustness to eavesdropping by nulling or limiting the wiretap channel when transmitting the data simultaneously to all the legitimate destinations.

5.4.3.2 Cooperative Relaying

When cooperative vehicles or pedestrians behave as relays in vehicular networks, they can employ amplify-and-forward (AF) or decode-and-forward (DF) protocol to forward the confidential information in a collaborative manner, such as cooperative beamforming and relay selection techniques. Cooperative beamforming is an efficient technique adopted at the relay nodes to forward confidential signals, which can provide both diversity and power gains for the destination to significantly

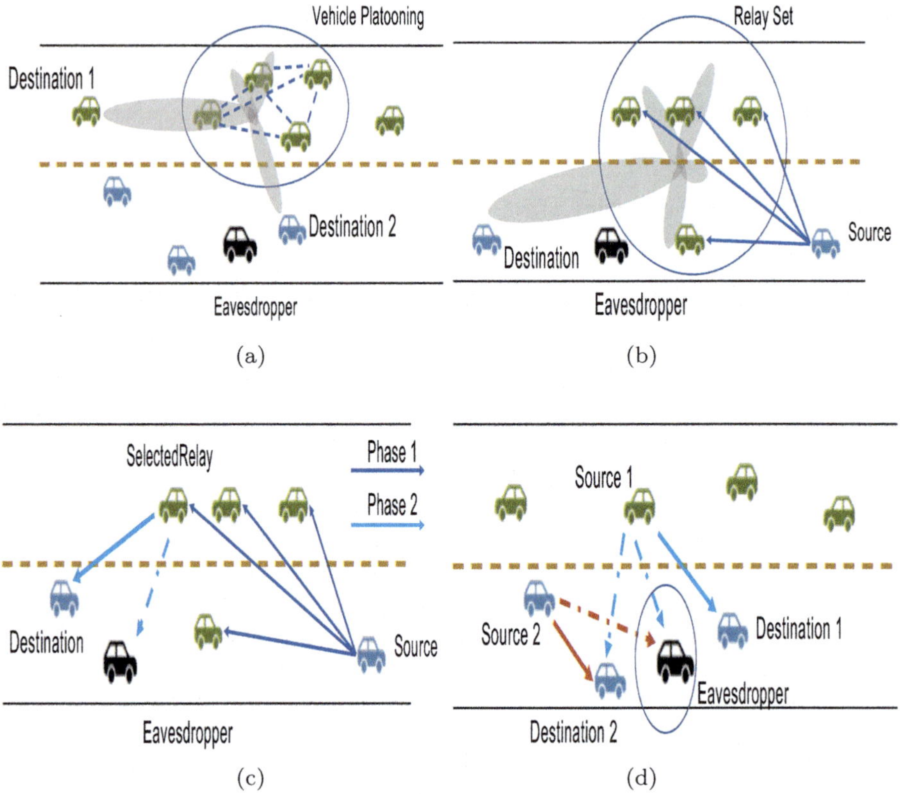

Fig. 5.17 Illustrated cooperative secrecy schemes in vehicular networks. (**a**) Cooperative MIMO through vehicle platooning. (**b**) Cooperative relaying: Cooperative beamforming. (**c**) Cooperative relaying: Relay selection. (**d**) Cooperative jamming based on resource sharing

enhance the rate of the legitimate channel. Concurrently, the forward signals can also be designed to superimpose destructively or even null out at the eavesdroppers. Another cooperative relaying scheme is to select a single relay that can yield the best secrecy performance for cooperation in the second phase, that is, relay selection. Intuitively, the node with the best channel to the destination and the worst channel to the eavesdropper should be selected as the active relay, which can increase the legitimate channel rate and decrease the wiretap channel rate simultaneously. It sacrifices secrecy performance for lower complexity since cooperative beamforming requires time and frequency synchronization, which results in extra overhead.

5.4.3.3 Cooperative Jamming

Cooperative jamming is also known as cooperative artificial noise transmission, or noise forwarding. Cooperative vehicles become jammers to transmit no-information-bearing signals (artificial noise) to cover confidential signals. Since the jamming signals will interfere with both destination and eavesdroppers, they should be designed carefully to establish the superiority of the equivalent legitimate channel to the equivalent wiretap channel in terms of SINR.

5.4.3.4 Hybrid Cooperative Secrecy Scheme

Hybrid cooperative secrecy schemes are proposed to combine the advantages of both the relaying and jamming strategies to further enhance secrecy capacity. In a hybrid relaying and jamming scheme, multiple cooperative nodes are divided into two groups: relay group and jammer group. The nodes in the relay group help relay confidential signals, while those in the jammer group perform cooperative jamming. In such a way, the legitimate channel is improved, and concurrently the eavesdropper is perturbed, leading to enhanced secrecy capacity.

5.4.4 Secure Routing for Multi-hop V2X Communications

In vehicular networks, multi-hop data transmission among vehicles is a common communication manner. It is a big challenge to achieve both efficient and secure routing for multi-hop V2X communications due to the high mobility of vehicles, the fast changing network topology, and the vulnerability of multi-hop transmissions. Taking the confidentiality and integrity into consideration, many secure routing protocols have been proposed for mobile ad hoc networks in the literature [37]. Almost all these existing protocols are designed by combining traditional ad hoc routing algorithms with high-layer cryptography methods or trust mechanisms. However, in vehicular networks, which is an open wireless environment and contains a huge number of entities therein, traditional encryption-based secure routing protocols will be too burdensome due to the secret keys distribution and updating among distributed moving vehicles. Physical layer security techniques can be employed into the secure routing design for multi-hop V2X communications to effectively improve the secrecy of the confidential data transmitted over multiple hops.

5.4.4.1 Secrecy-Integrated Routing Metric Design

In order to achieve both efficient and secure routing by exploiting physical layer security techniques, a novel routing metric by jointly taking the transmission efficiency and the secrecy performance is required as an optimization indicator in physical layer security-based secure routing protocol design. Then, the secrecy-integrated routing metric can be defined as $M_r = \mathcal{F}(C^s, D)$, where $\mathcal{F}(\cdot)$ denotes an appropriate utility function, C^s is one-hop secrecy capacity or achievable secrecy rate, and D is a simplified factor that indicates the total transmission delay in a qualitative manner, such as the number of potential transmission hops and the distance between two neighboring vehicles in a specific hop. For instance, a feasible secrecy-integrated routing metric can be given as

$$M_r = \left(C^s\right)^{\alpha} d^{\beta} \tag{5.36}$$

where C^s is the secrecy capacity for the next-hop transmission, d is the distance between the current vehicle and the selected next-hop relay, and α and β represents the weight factors, which can be dynamically adjusted according to real-time network states and practical requirements to achieve a good tradeoff between the transmission efficiency and secrecy performance. Note that the larger d is, the lower the total transmission delay from the source to the intended destination would be. Therefore, both α and β should be positive to guarantee that the utility function of the secrecy-integrated routing metric is monotone increasing with both C^s and d, in accord with our objective in achieving larger overall secrecy capacity and lower transmission delay through secure routing protocols.

5.4.4.2 Efficient Distributed Secure Routing Protocol

In this section, based on the designed secrecy-integrated routing metric given in (5.36), an efficient secrecy-based distributed secure routing (S-DSR) protocol is provided. Given that each vehicle knows both the direct distance and the CSI to its neighboring vehicles, the S-DSR protocol chooses the vehicle with the maximum secrecy-based routing metric as the relay for each transmission hop. Moreover, in order to guarantee positive overall secrecy capacity of the multi-hop data transmission, a wait and prediction scheme is also included in the S-DSR protocol to make sure that the achievable secrecy rate of each hop is positive.

Specifically, in the S-DSR protocol, the current vehicle with the data first calculates the secrecy-integrated routing metric of each active neighboring vehicle based on the collected distance and CSI. If there is at least one available neighboring vehicle that can achieve positive C^s, the vehicle with the highest secrecy-integrated routing metric will be selected as the relay for next hop. Otherwise, the current vehicle will wait for a pre-determined waiting time window CW_{min}, and predict the future routing metric of its neighboring vehicles at the end of the waiting duration based on the stored and updated information. The current vehicle will forward the data packets to the selected relay with the maximum routing metric when there is an available vehicle after the current waiting duration. If there is still no available vehicle achieving positive C^s, the waiting time window CW will be enlarged according to the binary exponential rule, that is, $CW = CW_{\mathrm{min}} \times 2^{n-1}$, where n is the number of the waiting durations that the current vehicle has undergone. The current vehicle as a relay would drop the transmitted data packets once the current total waiting time after the current waiting duration is larger than a pre-set valid relaying period. A failure feedback will be sent to the previous vehicle if the current vehicle fails to select an available vehicle to forward the data packets. And the previous vehicle needs to re-select a vehicle for data packets relaying.

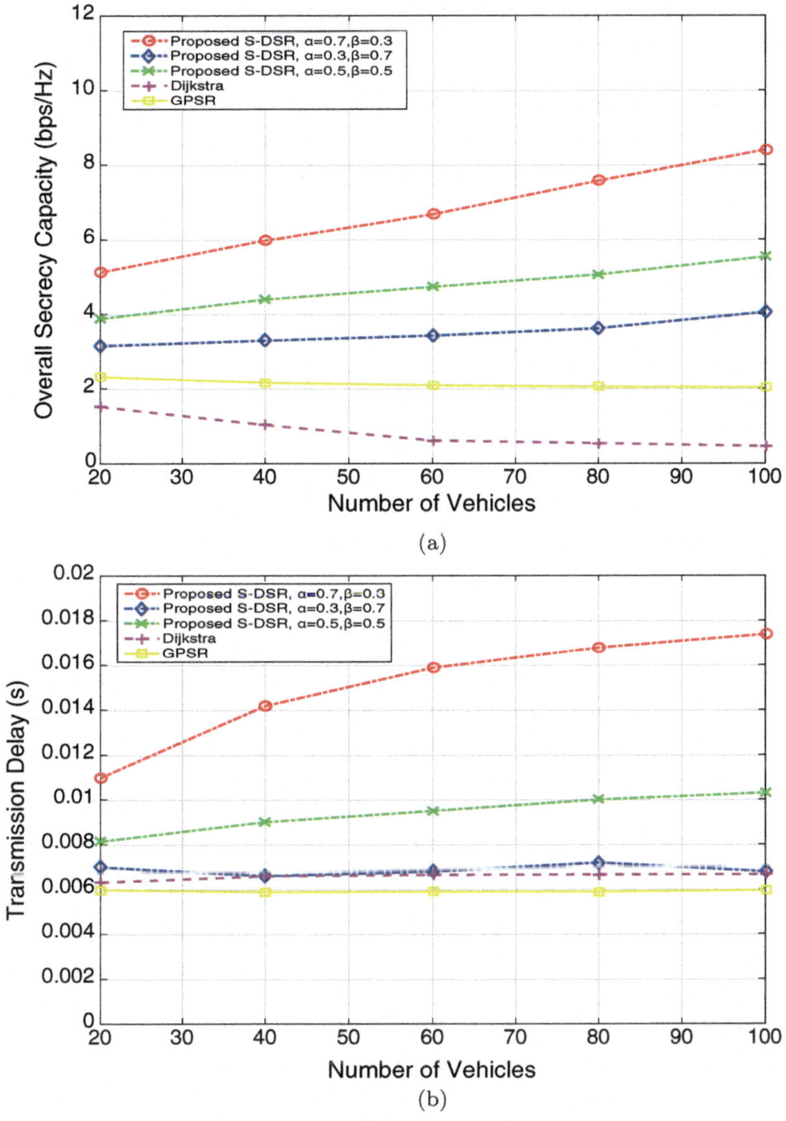

Fig. 5.18 Secure routing performance comparison in a 500 × 20 m simulated area where there is one moving eavesdropper and the effective transmission distance of each vehicle is 100 m. (**a**) Secrecy capacity performance comparison. (**b**) Transmission delay performance comparison

In Fig. 5.18, the efficiency of the S-DSR protocol is evaluated compared with the classical Dijkstra algorithm and the well-known greedy perimeter stateless routing (GPSR) protocol [38], in terms of both overall secrecy capacity and transmission delay performance. From Fig. 5.18, it can be found that when α in (5.36) increases, both the overall secrecy capacity and transmission delay increase, whereas higher value of β leads to both lower overall secrecy capacity and transmission delay. Therefore, there is a best tradeoff between α and β in achieving high overall secrecy capacity and low transmission delay simultaneously. In addition, it can be also seen that with a comparable transmission delay with the Dijkstra algorithm and the GPSR protocol (for instance when $\alpha = 0.3$ and $\beta = 0.7$ in the S-DSR protocol), the S-DSR protocol can achieve much higher secrecy performance.

5.5 The Next Leap

Currently, many essential vehicle-related applications are still facing big challenges, but integrated with enhanced vehicular communications and networking, promising solutions can be obtained in the near future.

5.5.1 Cooperative Sensing for Autonomous Driving

As an emerging technology attracting exponentially growing research and development interests from various sectors including academia, industry and government, autonomous driving is expected to bring numerous benefits to our everyday life, including increased safety, alleviation of traffic congestion, improved parking, and more efficient utilization of transportation resources [39, 40].

The first and foremost task of autonomous driving is environmental information acquisition. The current solution, however, almost exclusively relies on the limited view range of a single vehicle, which has been one main reason behind the multiple accidents involving intelligent vehicles lately [41]. To overcome the limitation of the current design strategy, cooperation among vehicles to achieve the situational awareness is inevitable. Such a cooperative framework requires autonomous vehicles to be capable of frequent and demanding exchange of data and information. Hence, this is a two-sided problem. On the one hand, vehicle cooperation strategies need to be designed subject to the capability of the wireless communication and networking. On the other hand, the wireless communication systems need to be adaptively designed to cope with the needs of cooperative vehicle sensing and driving, subject to the specific data rate and latency requirements. Moreover, on-vehicle storage and computing are additional issues that need to be jointly considered with the communication systems. We expect that these would require revolutionary progress at every level of the network stack.

5.5.2 *Storage and Processing of Huge Vehicular Data*

Typically, 5G-VCN framework has three levels of storage and processing units, namely, onboard, fog, and cloud [42]. Therefore, it is of great significance for the appropriate designation of the intelligent vehicles (IVs)'s data. According to the respective features of the onboard, fog, and cloud storage and processing units, we could delineate the IV data and accordingly allocate the tasks with stringent real-time constraints to the onboard units, the complex and latency-tolerant tasks to the cloud units, and tasks with intermediate requirements to the fog units. Thanks to the powerful processing capability of 5G-VCN at the fog and cloud side, complex data analytics that are difficult or even impossible for processors onboard the vehicle would be a realistic possibility. In the meantime, 5G-VCN can collectively combine the massive data not only from multiple IVs but also from human-operated vehicles. This will lead to multilateral benefits to the learning process of IVs, and thus enhanced reliability and safety of IVs, especially in the envisioned long period during which self-driving and human-intervened IVs coexist.

References

1. A. G. Boulanger, A. C. Chu, S. Maxx, and D. L. Waltz, "Vehicle electrification: Status and issues," *Proc. IEEE*, vol. 99, no. 6, pp. 1116–1138, Jun. 2011.
2. R. Zhang, X. Cheng, and L. Yang, "Energy management framework for electric vehicles in the smart grid: A three-party game," *IEEE Communications Magazine*, vol. 54, no. 12, pp. 93–101, Dec. 2016.
3. C. Liu, K. T. Chau, D. Wu, and S. Gao, "Opportunities and challenges of vehicle-to-home, vehicle-to-vehicle, and vehicle-to-grid technologies," *Proc. IEEE*, vol. 101, no. 11, pp. 2409–2427, Nov. 2013.
4. "The Energy Internet aka The Smart Grid - Putting It All Together," GreenAngel Energy Report.
5. A. Q. Huang, *et al.*, "The future renewable electric energy delivery and management (FREEDM) system: The energy Internet," *Proc. IEEE*, vol. 99, no. 1, pp. 133–147, Jan. 2011.
6. J. A. P. Lopes, F. J. Soares, and P. M. R. Almeida, "Integration of electric vehicles in the electric power system," *Proc. IEEE*, vol. 99, no. 1, pp. 168–183, Jan. 2011.
7. P. Palensky and D. Dietrich, "Demand side management: Demand response, intelligent energy systems, and smart loads," *IEEE Transactions on Industrial Informatics*, vol. 7, no. 3, pp. 381–388, Aug. 2011.
8. H. K. Nguyen and J. B. Song, "Optimal charging and discharging for multiple PHEVs with demand side management in vehicle-to-building," *Journal of Communications and Networks*, vol. 14, no. 6, pp. 662–671, Dec. 2012.
9. R. Yu, J. Ding, W. Zhong, Y. Liu, and S. Xie, "PHEV charging and discharging cooperation in V2G networks: A coalition game approach," *IEEE Internet of Things Journal*, vol. 1, no. 6, pp. 578–589, Dec. 2014.
10. Z. Tan, P. Yang, and A. Nehorai, "An optimal and distributed demand response strategy with electric vehicles in the smart grid," *IEEE Transactions on Smart Grid*, vol. 5, no. 2, pp. 861–869, Mar. 2014.

11. S. Bashash and H. K. Fathy, "Cost-optimal charging of plug-in hybrid electric vehicles under time-varying electricity price signals," *IEEE Transactions on Intelligent Transportation Systems*, vol. 15, no. 5, pp. 1958–1968, Oct. 2014.
12. X. Cheng, R. Zhang, and L. Yang, "Consumer-centered energy system for electric vehicles and the smart grid," *IEEE Intelligent Systems*, vol. 31, no. 3, pp. 97–101, May 2016.
13. X. Cheng, *et al.*, "Electrified vehicles and the smart grid: the ITS perspective," *IEEE Transactions on Intelligent Transportation Systems*, vol. 15, no. 4, pp. 1388–1404, Aug. 2014.
14. M. Wang, R. Zhang, and X. Shen, Mobile Electric Vehicles: Online Charging and Discharging, Springer, 2016.
15. R. Zhang, X. Cheng, and L. Yang, "Flexible energy management protocol for cooperative EV-to-EV charging," in *Proc. IEEE GLOBECOM'16*, Washington, D.C., USA, Dec. 2016.
16. R. Zhang, X. Cheng, and L. Yang, "Stable matching based cooperative V2V charging mechanism for electric vehicles, in *Proc. IEEE VTC 2017-Fall*, Toronto, Canada, Sept. 2017.
17. R. Zhang, X. Cheng, and L. Yang, "Flexible energy management protocol for cooperative EV-to-EV charging," *IEEE Transactions on Intelligent Transportation Systems*, 2018, to appear.
18. S. Beer *et al.*, An economic analysis of used electric vehicle batteries integrated into commercial building microgrids, IEEE Transactions on Smart Grid, vol. 3, no. 1, pp. 517–525, Mar. 2012.
19. D. Gale and L. S. Shapley, "College admissions and the stability of marriage," American Mathematical Monthly, vol. 69, no. 1, pp. 9–15, Jan. 1962.
20. B. Ahlgren, C. Dannewitz, C. Imbrenda, D. Kutscher, and B. Ohlman, "A survey of information-centric networking," *IEEE Communications Magazine*, vol. 50, no. 7, Jul. 2012.
21. D. Malak and M. Al-Shalash, "Optimal caching for device-to-device content distribution in 5G networks," in *Proceedings of IEEE Globecom Workshops (GC Wkshps)*, Austin, TX, USA, Dec. 8–12, 2014, pp. 863–868.
22. B. Hu, L. Fang, X. Cheng, and L. Yang, "In-Vehicle Caching (IV-Cache) via Dynamic Distributed Storage Relay (D^2SR) in Vehicular Networks," submitted to IEEE Transactions on Vehicular Technology.
23. A. G. Dimakis, K. Ramchandran, Y. Wu, and C. Suh, "A survey on network codes for distributed storage," *Proceedings of the IEEE*, vol. 99, no. 3, pp. 476–489, 2011.
24. B. Hu, L. Fang, X. Cheng, and L. Yang, "Vehicle-to-Vehicle Distributed Storage in Vehicular Networks," in *Proceedings of IEEE International Conference on Communications (ICC)*, Kansas City, MO, USA, May 20–24, 2018.
25. K. Abboud and W. Zhuang, "Stochastic analysis of a single-hop communication link in vehicular ad hoc networks," *IEEE Transactions on Intelligent Transportation Systems*, vol. 15, no. 5, pp. 2297–2307, Oct. 2014.
26. R. Ding, T. Wang, L. Song, Z. Han, and J. Wu, "Roadside-unit caching in vehicular ad hoc networks for efficient popular content delivery," in *Proceedings of IEEE Wireless Communications and Networking Conference (WCNC)*, New Orleans, LA, USA, Mar. 9–12, 2015, pp. 1207–1212.
27. M. Sankaran, "On the non-central chi-square distribution," *Biometrika*, vol. 46, no. 1/2, pp. 235–237, 1959.
28. N. Golrezaei, A. G. Dimakis, and A. F. Molisch, "Device-to-device collaboration through distributed storage," in *Proceedings of IEEE Global Communications Conference (GLOBECOM)*, Anaheim, CA, USA, Dec. 3–7, 2012, pp. 2397–2402.
29. L. Idir, S. Paris, and F. Naït-Abdesselam, "Optimal caching of encoded data for content distribution in vehicular networks," in *Proceedings of IEEE International Conference on Communication Workshop (ICCW)*, London, UK, Jun. 8–12, 2015, pp. 2483–2488.
30. Z. Hu, Z. Zheng, T. Wang, L. Song, and X. Li, "Roadside unit caching: Auction-based storage allocation for multiple content providers," *IEEE Transactions on Wireless Communications*, vol. 16, no. 10, pp. 6321–6334, Oct. 2017.
31. Y. Zou, J. Zhu, X. Wang, and L. Hanzo, "A survey on wireless security: Technical challenges, recent advances, and future trends," *Proc. the IEEE*, vol. 104, no. 9, pp. 1727–1765, Sept. 2016.

32. R. Zhang, X. Cheng, and L. Yang, "Cooperation via spectrum sharing for physical layer security in device-to-device communications underlaying cellular networks," *IEEE Transactions on Wireless Communications*, vol. 15, no. 8, pp. 5651–5663, Aug. 2016.
33. R. Zhang, L. Song, Z. Han and B. Jiao, "Physical layer security for two-way untrusted relaying with friendly jammers," *IEEE Transactions on Vehicular Technology*, vol. 61, no. 8, pp. 3693–3704, Oct. 2012.
34. H.-M. Wang and X.-G. Xia, "Enhancing wireless secrecy via cooperation: Signal design and optimization," *IEEE Communications Magazine*, vol. 53, no. 12, pp. 47–53, Dec. 2015.
35. J. Chen, R. Zhang, L. Song, Z. Han and B. Jiao, "Joint Relay and Jammer Selection for Secure Two-Way Relay Networks," *IEEE Transactions on Information Forensics and Security*, vol. 7, no. 1, pp. 310–320, Feb. 2012.
36. L. Dong, Z. Han, A. P. Petropulu, and H. V. Poor, "Improving Wireless Physical Layer Security via Cooperating Relays," *IEEE Transactions on Signal Processing*, vol. 58, no. 3, pp. 1875–1888, Mar. 2010.
37. L. Abusalah, A. Khokhar and M. Guizani, "A survey of secure mobile ad hoc routing protocols," *IEEE Communications Surveys & Tutorials*, vol. 10, no. 4, pp. 78–93, Fourth Quarter 2008.
38. B. Karp and H. T. Kung, "GPSR: Greedy perimeter stateless routing for wireless networks," in *Proc. the 6th ACM International Conference on Mobile Computing and Networking (MobiCom'00)*, New York, NY, USA.
39. T. Luettel, M. Himmelsbach, and H. J. Wuensche, "Autonomous Ground Vehicles Concepts and a Path to the Future," *Proceedings of the IEEE*, vol. 100, no. Special Centennial Issue, pp. 1831–1839, May 2012.
40. G. Bresson, Z. Alsayed, L. Yu, and S. Glaser, "Simultaneous Localization and Mapping: A Survey of Current Trends in Autonomous Driving," *IEEE Transactions on Intelligent Vehicles*, vol. 2, no. 3, pp. 194–220, Sept. 2017.
41. S. W. Kim, W. Liu, M. H. Ang, E. Frazzoli, and D. Rus, "The Impact of Cooperative Perception on Decision Making and Planning of Autonomous Vehicles," *IEEE Intelligent Transportation Systems Magazine*, vol. 7, no. 3, pp. 39–50, Fall 2015.
42. X. Cheng, C. Chen, W. Zhang and Y. Yang, "5G-Enabled Cooperative Intelligent Vehicular (5GenCIV) Framework: When Benz Meets Marconi," *IEEE Intelligent Systems*, vol. 32, no. 3, pp. 53–59, May-June 2017.

Printed by Printforce, the Netherlands